THE REENGINEERING ALTERNATIVE

A Plan for Making Your Current Culture Work

THE REENGINEERING ALTERNATIVE

A Plan for Making Your Current Culture Work

William E. Schneider

IRWIN
Professional Publishing
Burr Ridge, Illinois
New York, New York

Senior sponsoring editor:	Cynthia A. Zigmund
Project editor:	Paula M. Buschman
Production manager:	Laurie Kersch
Designer:	Larry J. Cope
Art manager:	Kim Meriwether
Compositor:	BookMasters, Inc.
Typeface:	11/13 Palatino
Printer:	Book Press

Library of Congress Cataloging-in-Publication Data

Schneider, William E.
   The reengineering alternative : a plan for making your current culture work / William E. Schneider.
      p.   cm.
   Includes bibliographical references and index.
   ISBN 0-7863-0120-1
   1. Corporate culture.   2. Organizational effectiveness.
I. Title.
HD58.7.S344   1994
302.3′5—dc20   93–38860

Printed in the United States of America
              0 BP 1 0 9 8

To Kristine, Bill, and Michael

Preface

My colleagues and I utilized the concepts and methods discussed in this book in working with the senior executives of a large insurance company, our client. The company is composed of six subsidiary insurance companies. We surveyed all six to determine the core culture that existed in each of them. The survey indicated that four of the six companies were control cultures, one was changing from a control culture to a competence culture, and one was changing from a control culture to another culture yet to be determined. The CEO responsible for all six companies has been in his position for more than three years; he was in charge of the company moving from a control culture to a competence culture and was strongly inclined to move the other five companies in that direction. The survey and interview data revealed that this company was engaged in a struggle between two core cultures: the control culture and the competence culture.

We started our feedback and action planning session by asking these senior executives to provide us with individual lists of emotionally charged words that they found were prevalent in their relationships with one another, as individuals and as companies.

The sequence is important. The senior executives provided their lists of emotionally charged words *before* they were given the survey results and *before* they knew anything about core culture concepts. The individual lists were compiled into a single list.

After presenting the survey results, discussing the meaning of four core cultures, and discussing in depth the meaning of the control and competence cultures, we looked at the combined list of emotionally charged words. Excluding technical or product-specific words, *75 percent* of the emotionally charged words on the combined list were related to the control and competence cultures! The word "*control*" itself was mentioned eight times.

This book will give you an emotionally charged experience. You will see organizations with which you have been involved in a wholly different and clearer light. You will understand why organizations operate the way they do and how they are different from one another. You may discover for the first time exactly what kind of organization most fits you. You will, very likely, see yourself more clearly. You will better understand why you were unhappy in one organization and happy in another. You will gain insights on quick and effective strategies for managing organizations or working within them.

BACKGROUND

The premises put forth in this book provide perspectives on organizational culture and effectiveness that can be adopted and put to practical use. These premises are formulated from experience-based observations and the literature, primarily business literature, that describes cultural life within organizations. They spring from my practice as a consulting organizational psychologist over the last 20 years. The traditional academic model moves from theory to practice. This book, in contrast, develops from practice to theory and then moves back to practice.

This work is intended to be organic: it not only should prove useful to the reader in its present form but also should stimulate discourse and experimentation that will modify and enrich it.

WHO SHOULD READ THIS BOOK

Leaders and managers from any kind of organization should find this book stimulating, clarifying, integrating, and challenging. They will discover numerous implications from its contents that will suggest ways to enhance the success and effectiveness of their own organization. Chapter Eight (A Framework for Developing Your Organization) will help them get started on a program of organizational development.

This book is relevant for any executive who anticipates a merger with or acquisition of another organization. A prior understanding of the core culture will go a long way toward helping parties decide how viable a potential merger or acquisition might be, strategic and financial considerations notwithstanding. Should they decide to go ahead, this information will help to formulate an effective integration strategy. The material on core cultures will help those already part of a merger or acquisition to understand where the two formerly separate organizations fit naturally with one another, where they are misaligned, and what can be done to bring about more integration.

Executives anticipating a strategic alliance with another organization will find the concepts in this book of considerable value. (A strategic alliance is a *partial*, usually product-specific alliance between two companies—such as IBM & Apple's alliance a year ago. The *companies* remain separate.) Understanding your organization's core culture and that of a potential ally will help the alliance go much more smoothly and effectively, and provide a clear strategy for forging the alliance.

Leaders of multinational organizations will get a fresh and exciting perspective on their overall organization, subsidiary organizations and operations, and organizations and operations in other parts of the globe. Preliminary research suggests a broad correlation between the four organizational cultures discussed in this book and national cultures. At a minimum, these leaders will begin to understand important similarities and differences between the different cultural elements within their multinational organization.

Employees will find it helpful to learn which core culture characterizes their own organization and how that core culture explains why and how their organization does what it does. They will find it much easier to get their organizational bearings.

Consultants to organizations will discover a fresh, integrated way of looking at the organizations they assist. Each core culture has its own set of inherent strengths on which consultants can focus and encourage within their client organizations.

Professionals in charge of development will learn which core culture is present within their own organization and can focus their development efforts. Much in this book bears on recruiting, organizational structuring, management development, and organizational development.

Students enrolled in programs on organizational behavior will gain a unique perspective on the role of culture.

Finally, academicians and researchers might see opportunities to develop new theories and conduct empirical investigations that expand our understanding of organizational culture and change.

HOW TO READ THIS BOOK

You will find it helpful, and perhaps more interesting, to work your way through this book by reading first the Introduction section and then Chapter 1, the Meaning and Importance of Organizational Culture. This chapter clarifies and defines culture and provides you with a rationale for its importance.

Next, you should read Chapter 2 and complete the Core Culture Questionnaire. The questionnaire is designed to help you determine the core culture that exists within your organization.

When you have scored the questionnaire and determined which core culture exists within your organization, go directly to the chapter that discusses this kind of culture. You can see to what extent this book's depiction of your core culture rings true to you.

Next, it is recommended that you read Chapter 7 (The Genesis of Organizational Culture) and Chapter 8 (A Framework for Developing Your Organization). These two chapters provide a perspective on all four cultures together and propose a way to enhance the success of your organization by building on the strengths of its core culture.

Finally, delve into the material on the other three core cultures as your curiosity and interests dictate.

Acknowledgements

I am indebted to the clients with whom I have worked for more than 20 years. Their teachings about themselves and their organizations serve as a substrate for the central ideas in this book. I hope I am returning the favor by offering them these ideas for their consideration.

I particularly thank John Williams, client and friend, who graciously took time from his hectic schedule to provide a thoughtful critique of an earlier draft of this book.

Jeff McDaniel has been a wise and resourceful advisor throughout the process of getting this book published. Cindy Zigmund, senior editor at Irwin Professional Publishing, has been a very competent guide and co-strategist.

I owe a number of colleagues a special debt of gratitude. Nick Colarelli, who stimulated my interest in organizations and organizational behavior, reviewed my manuscript and provided me with many valuable insights. Paul White helped me put the concepts of this book into the larger perspective. Rad Eanes gave graciously of his time and added considerably to my thinking about organizational behavior and organizational change. Terry Tutchings not only helped substantially reorganize this book but also competently edited the entire text and did much to help me put the material into a larger perspective.

The colleagues to whom I am also indebted are my former co-workers at Somerville & Company, Inc.: Kevin Somerville, Diane Hill, Steve Bloom, Diane Bradford, Guy Cornelius, Andy Selig, Sally Sporer, and Jon Ziarnik. They have taken much time from busy schedules to review the manuscript, offer helpful insights, and recommend changes. I will forever be grateful for their support and encouragement.

Diana Dehn, Patti Eisenbrandt, and Marge Smith were immeasurably helpful in typing and preparing the manuscript. Diana

Dehn, particularly, patiently worked and reworked various parts of the manuscript.

This book would not have been possible without the support, dedication, and good judgment of my wife, Kristine. She persisted with me, even when the work at times seemed insurmountable. She was always there when I needed to discuss something about the book. Mostly, I appreciate her willingness to sacrifice as much as she did so that I could complete the project.

Finally, I want to thank my two sons, Bill and Michael, for their encouragement, interest, and care. They, too, made sacrifices so that I could complete the writing of this book.

Contents

Introduction

SOS, Inc.,[1] a wholly owned subsidiary of a large East Coast newspaper publishing company, is a software development company based in the Midwest. SOS provides custom software development for small- to medium-sized newspapers. By 1987, its revenues were approximately $21 million and the company was very profitable. SOS's principal way of doing business was to form partnerships with its customers to build highly customized software for the various functions of its client newspapers. SOS saw each client newspaper as unique and expressly tailored each software package to the client's needs. SOS software engineers spent much of their time on client premises where, on many occasions, they could be found in front of a computer with people from the client organization working on a software problem. Inside SOS, management and staff functioned as a team or series of teams. All important matters were decided by these teams, particularly the top management team.

In 1988, the parent company of SOS appointed a new president of SOS. Ellen B. was very competent and took hold of her job with great energy and drive. She soon mandated an important addition to the focus of SOS's business. Besides providing custom software development, SOS also would develop and market standardized software for newspapers. SOS engineers were required to do both kinds of work—custom and standardized software development. Ellen also declared that SOS was to focus on being an "excellent" company and that the managers and staff were to achieve the best, state-of-the-art, standardized software possible for the complete newspaper market.

Three years later, in 1991, SOS's revenues were down 50 percent to $10.5 million, the company was operating in the red, morale was low, half the staff had left, and productivity was well below that of earlier years. What happened to SOS?

Let's find out first what happened to Change Technologies, Inc., another company, once a thriving and spirited management consulting organization staffed with creative and committed people, which ran into difficulty. Between 1970 and 1974, its revenues grew to $4 million—roughly doubling each year of that four-year period. The firm was loosely organized and employees were given every opportunity for self-expression and experimentation. Commitment was high. People often worked late into the night on new services and new products. Considerable attention was paid to the purposes of the organization and to the purposes deeply aspired to by the firm's clients. Clients were pleased with Change Technologies' services. In only four years, the firm developed seven new services, four new programs, and two new products (change instruments). Creativity reigned supreme. The firm seemed to have everything going for it.

But by the end of 1975, Change Technologies was out of business. First the president left, then the executive vice president. One by one, staff departed. By September 1975, the firm consisted of one person who shut things down.

What happened at Change Technologies? Around the end of 1974 and for much of 1975, things got out of control. People started going off on individual tangents. Strife set in. Staff complained of being lost and confused. Misunderstanding and ill will prevailed. No one could hold things together. Almost overnight the organization dissolved.

One more situation. Sigma Electric is a large electric utility company headquartered in a large city in the South. Its revenues were about $1.5 billion in 1991, and its profits had been moving steadily higher until 1987, when they began to drift lower (where they remain to this day). In 1985, Sigma's longtime CEO, Robert L., retired and was replaced by John S., who had been senior vice president of finance. John S. was very different from his predecessor. Robert L. had been a definitive and authoritative leader who provided clear and firm direction for Sigma. Robert and his predecessors had built an organization that was functionally driven, hierarchical, and dominant in its marketplace. John S.'s approach as CEO, by contrast, has been reserved, thoughtful, and deliberate. His emphasis has been much more hands-off and relaxed. By all accounts, the leadership change has not worked.

Sigma's performance has faltered. Management and employees feel adrift. Decisions are slow in coming from the top. Information flows upward within the company, but little communication comes back down. People within the company believe that Sigma Electric has lost its direction and is losing its dominant place in its marketplace.

What happened to these organizations? What caused them to run into trouble?

First, each organization had a core culture, a fundamental way of accomplishing success, that the employees knew nothing about. Briefly, a control culture has to do with power, a collaboration culture is all about teams and teamwork, a competence culture focuses on achievement, and the cultivation culture is concerned with growth and potential. In the three examples, management and staff did not know their organization's central nature. They also did not know how important the core culture is to the organization's foundation. This lack of insight caused them to do things with each organization that they would never have done otherwise.

A second reason that each of these organizations faltered is that each, in its own way, lacked "balanced integrity," or organizational effectiveness.

SOS, Inc., faltered because Ellen B. tried to forcefully integrate one core culture, a competence culture, into another, a collaboration culture. Prior to Ellen B.'s arrival, SOS was functioning successfully as a collaboration culture. Not seeing this nor believing in the value and power of such a core culture, she blindly tried to force two very different worlds to function in an integrated manner. It didn't work and it could never have worked. Unwittingly, she built in a serious misalignment the minute she announced that SOS was going into the standardized software business and that people were to begin developing state-of-the-art products. With the best of intentions, she had asked her organization to take on and live out two very different—indeed, inherently contradictory—personalities within the same "body."

Change Technologies faltered because it let some of the key strengths of its core culture get out of balance. Partly blinded by its success, the company began operating in the extreme. Change Technologies was a cultivation culture. Reasonable self-expression turned into self-aggrandizement. Intentionality prevailed over everything else, another extreme state of affairs that only served to help people avoid accountabilities. Finally, the leadership of the firm went too far by empowering and cultivating the staff without channeling the staff's efforts toward fulfilling the purposes of the firm and of the firm's clients.

Sigma Electric's difficulties illustrate how incompleteness can get a company into trouble. Sigma is a control culture, which needs a firm, directive, and authoritative leadership approach in order to be complete. John S.'s leadership leaves a big gap within the organization. As a control culture, Sigma is incomplete unless it is operated in an authoritative manner. Its culture lacks an important element that it needs to operate effectively.

FOCUS, INTEGRATION, BALANCE, AND COMPLETENESS

This book is based on the premise that nothing is so practical as a good theory—that is, if you get it theoretically right, in the first place, then everything practical to be gained falls automatically into place.

The material in this book applies to organizations of any kind: large, medium, or small; profit and nonprofit; product-driven and service-driven. It applies to organizations that are still just a twinkle in someone's eye.

Successful organizations are focused. SOS, Change Technologies, and Sigma Electric were out of focus because they were unaware of their internal, core nature. Lacking critical self-insight into their core culture, they headed off in directions or allowed changes to take place that were inherently doomed from the start.

An organization's core culture is a fundamentally important element to that organization's focus. Any organization that has been functioning in a reasonably healthy manner has embedded within it one of the four possible core cultures: the "control" culture, the "collaboration" culture, the "competence" culture, and the "cultivation" culture.[2]

Focus also has to do with living out one's organizational destiny. Assuming that an organization has adopted a core culture that is inherently congruent with the nature of that organization's enterprise and environment (an assumption that does not always hold true), that organization succeeds by building on and working with that core culture. Each organization's cultural nature and unique mix of cultural strengths and weaknesses frames the essential basis for what that organization can become. Organizational success and the essential paradigm for getting there lie within.

This premise stands out in marked contrast to much of the management and organizational literature written in the last decade. A plethora of books and articles encourage, directly or indirectly, readers to look *outside* themselves and their organizations for direction, formulas, and lessons for success. Readers are encouraged to adopt the practices of the "excellent" companies, the Japanese, the "innovators," or the "changers," those who are the most intense, and those who emphasize "transcendence" and purpose. While these proposals may have considerable value, they are often unwittingly pulling their readers off center by insisting that conclusions reached from observing the practices of one kind of organization or group of organizations must apply to all organizations. This is surely an overgeneralization or a kind of organizational benchmarking writ large.

Assuming that your organization's core culture is congruent with the nature of your enterprise, you must first turn inside to increase the success of your organization. There you can identify the core culture, the natural definition of your success, the natural approach to your customers and constituents, the natural leadership focus, and many other parts of your organization's nature. This book also postulates that, given the nature of your core culture, you have a wellspring of strengths to build on and a set of weaknesses that you can either minimize or compensate for. In addition, this book proposes that whatever is being touted for you to adopt for your own organization is already operative in one of the four core cultures; therefore, knowing which core culture guides your organization is the key place to start. Trying to transplant elements from one or all of the other three core cultures into your core culture will, more often than not, lead to their rejection unless you consider such actions carefully.

Successful organizations are well-integrated. Ellen B. tried forcefully to integrate the essence of a competence culture into the existing collaboration culture of SOS. Her initiative didn't work. Instead of making the company grow, Ellen took it backward. The forced integration had little chance of working because the competence culture is the diametric opposite of the collaboration culture.

Integration has to do with alignment and fit. Successful organizations are internally coherent and congruent. The different parts operate in sync with one another.

Let's return to a point made earlier: you cannot force-feed a generalized principle, practice, or program into any one core culture and expect it to work unless it is congruent with the core culture of the organization. Some consultants do not like to broach this key issue. Many total quality management (TQM) programs available today for organizations are a compilation of elements from the collaboration, competence, and cultivation cultures. This is a problem in itself, yet these TQM programs are touted as applicable for all kinds of organizations. This presents a considerable integration problem. Because each core culture is unique and built a certain way, to propose that a blanket TQM program will effectively integrate with all four core cultures is probably incorrect. A recent study by Ernst & Young and the American Quality Foundation[3] found that the "total quality movement, one of the biggest fads in corporate management, is floundering . . . Despite plenty of talk and much action, many American companies are stumbling in their implementation of quality-improvement efforts, . . . Many quality-management plans are simply too amorphous to generate better products and services." This comprehensive study involved 584 companies in the United States, Canada, Germany, and Japan. This finding is not surprising given the importance of integration and alignment.

Successful organizations are balanced. Change Technologies began operating in the extreme and eventually went out of existence. An inherent strength and an integral element to this organization's cultivation culture was allowed to run unfettered and unchanneled, eventually contributing mightily to the organization's decline. Balanced organizations operate in a state of equilibrium. They build in procedures to ensure that things do not get out of hand. They also build in elements from other core cultures, particularly opposite core cultures, that are needed to keep the organization in a state of equilibrium.

Finally, successful organizations are close to complete. They have no glaring holes where important elements necessary for effectiveness are missing. The management style of Sigma Electric's John S. caused the whole organization to be too incomplete. The leadership approach needed by Sigma is missing and the organization suffers as a result.

Completeness means that all or almost all of the important elements necessary for an organization's success are present and functioning.

A NATURAL TOPOGRAPHY OF ORGANIZATIONS

The focus on culture in this book allows for a topographical analysis of organizations that helps integrate important organizational phenomena that often have been treated separately.

For example, a great deal has been written on organizational leadership, much of it oriented toward identifying what makes leaders effective and successful. This book will connect the four kinds of leaders—director, standard setter, participative, charismatic—to the four core cultures. The leader as director is naturally congruent with the control culture. The participative leader fits hand-in-glove with the collaboration culture. The leader as standard setter operates naturally within the competence culture. The charismatic leader aligns naturally with the cultivation culture. Considered from the perspective of the natural fit between the kind of leader and the kind of core culture, the notion of leadership effectiveness takes on a different and critical dimension. To an important extent, leadership effectiveness is a question of degree of fit. A leader's effectiveness can be measured by the degree the leader's approach is integrated with the organization's core culture.

Mergers and acquisitions are another example. When they fail, as is often the case, they do so in large measure because the leadership of one core culture has tried to force fit its core culture into a different core culture. The result is ongoing frustration, conflict, loss of productivity, and reduced profitability.

The reader will discover that each core culture has its own brand of learning, knowing, and deciding. The control culture emphasizes "organizational systematism" when coming to belief (knowledge) and making decisions. The collaboration culture relies on "experiential knowing." The competence culture keeps its focus on "conceptual systematism." And, the cultivation culture emphasizes "evaluational knowing." All of this has deep implications for organizational planning, strategizing, and direction setting, among other matters. And each core culture takes a different approach to the development of its employees, supervisors, and managers. The control culture emphasizes didactic and structured development programs. The collaboration culture emphasizes group experiences. The competence culture places priority on the development of specialized knowledge and skills. The cultivation culture typically has its people in a developmental mode much of the time that people are at work.

There are many more patterns. Each core culture operates differently when it comes to additional areas such as promotion practices, recruiting and hiring, the use of power and authority, structuring, approaching customers or constituents, management style, task focus, the role of the employee, and approach to change.

NOTES

1. SOS, Inc., Change Technologies, Inc., and Sigma Electric are pseudonyms for actual organizations. The events depicted actually occurred.
2. The words that characterize the four core cultures—control, collaboration, competence, and cultivation—were selected because each captures the *essence* of one of the four cultures. The fact that each begins with the letter *c* is happenstance. No positive or negative meanings are conveyed by these words, nor is it implied that one word is better than another. Words used to describe one culture may also have elements that describe or apply to another. For example, "competence"; each culture needs competent people to do its work. The use of competence to describe a culture does not mean that only the competence culture needs competent people. Each word is purely an attempt to be neutrally, but cogently, descriptive.
3. G. Fuchsberg, "Quality Programs Show Shoddy Results," *The Wall Street Journal*, May 14, 1992, sec. B.

Chapter One

The Meaning and Importance of Organizational Culture

Organizational culture[1] is a much-discussed topic today and will soon emerge as a pivotal frame of reference for every leader or manager in any organization.

Definitions of culture abound from the esoteric "a system of symbols" to the incomprehensible "a bubble of meaning covering the world." In this book, organizational culture is an organization's essential way to success. Every business, church, school, or symphony orchestra seeks its own brand of prosperity, its brand of success. Every organization formulates and implements its own essential way to get there. The essential way, or fundamental method of operation, chosen by an organization establishes and quickly becomes equivalent to the culture of that organization.

Organizational culture may be defined as the way we do things around here in order to succeed.[2]

THE MEANING OF CORE CULTURE

The concept of *core* culture means the innermost part of an organization's culture. It is the nucleus of the culture.

As one moves from the nucleus to the periphery, however, one finds a myriad of different-looking organizations. Indeed, a single organization can exhibit characteristics of all four core cultures. There is no intent to oversimplify the nature of organizations. They come in all shapes and sizes.

An analogy from botany helps explain the nature of organizations. Botany, the science of plants, has so far identified about 300,000 different species of plants in the world. While the existence of 300,000 species of organizations is unlikely, the botanical concept of species corresponds to the concept of organization. As one goes deeper into plant life, however, the *form* and structure of plants narrows to subkingdoms. In our analogy, core culture corresponds to the subkingdom.

In addition, core culture conveys no meaning of better or higher or superior. One core culture is not better than another. Each has its own mix of strengths and weaknesses. Each has its own role to play in the structure and conduct of organizational life. Core culture is a naturalistic concept; like individual character, it develops naturally in the course of human events. It is possible that one core culture is more suited naturally to one kind of endeavor than another, but that thesis is beyond the scope of this book.[3]

CULTURE AND LEADERSHIP

Organizational culture is intimately tied to leadership. How the leaders of an organization believe things should be done drives the kind of culture that is established.[4]

Leaders build paradigms about how things should be. Thomas Kuhn defines *paradigm* as a "constellation of concepts, values, perceptions, and practices shared by a community which forms a particular vision of reality that is the basis of the way a community organizes itself."[5] In a more general sense, a paradigm is the way people, including leaders, make sense of their world. Paradigms give a picture of how things ought to be overall. And they derive from how we are socialized to believe the world ought to be—including how to run a work organization. Leaders establish their organizational cultures to fit their personal paradigms. This process operates at a societal level (e.g., the US Constitution) as well as the organizational level. It is a fundamental underpinning to the formation of an organization's culture.

Where do leaders get their organization's cultural paradigms? They get them from the value they place on their own individual socialization experience(s) and from their own individual (and primary) motive system.

Leadership cultural paradigms have their fundamental base in one of four social institutions and one of four correlative individual motives. The four social institutions are the

- Military
- Family and/or athletic team
- University
- Religious Institutions (*Church, synagogue, or mosque*)[6]

The four core cultures trace their origin to these four social institutions.

Leaders also bring their essentially individual motives to the task of forming and organizing an organization. Four essential individual motives closely parallel the four social institutions:

- Power
- Affiliation
- Achievement
- Growth, or self-actualization[7]

The motive of power fits naturally with the military, the affiliation motive ties in with the family and/or athletic team, the motive of achievement coincides with the university, and the motive of growth fits naturally with religious institutions.

This natural and striking parallel between the four fundamental social institutions and the four basic individual motives speaks to their essential role as paradigms for forming and building organizations. When combined, each social institution and individual motive serve as a *leadership* paradigm for culture formation. They also serve as the basis for the four core cultures of organizations.

Business school theorists insist that a culture is essentially formed from what it takes for a particular business organization to succeed in its marketplace.

Each company faces a different reality in the marketplace depending on its products, competitors, customers, technologies, government influences, and so on. To succeed in its marketplace, each company must carry out certain kinds of activities very well. In some markets that means selling; in others, invention; in still others, management of costs. In short, the environment in which a company operates determines what it must do to be a success. This business environment is the single greatest influence in shaping a corporate culture.[8]

While the nature of an organization's business environment and what it takes to succeed (in that environment) are clearly very important, the single greatest influence in shaping any culture is the organization's leadership contingent and, most important, how those leaders perceive that business environment and what it takes to succeed in it. An essential ingredient is how leaders believe an organization should be modeled and how it ought to be run. This explains why different organizations within the same industry are run so differently from one another; for example, Delta Airlines vs. Continental Airlines. If success in the business environment is the principal contributor to culture formation, then it is reasonable to predict that all companies in the same industry would have the same culture. However, while same-industry companies have considerable functional overlap, they often differ greatly in their organizational cultures. This fact strongly suggests that leadership is the principal factor in culture formation and that the leader's paradigm for organizing the company is more fundamental than anything else. Hal Geneen, former CEO of the ITT Corporation, said that "all organizations, large and small . . . reflect the personality and character of the man or men who lead them. The chief executive establishes the personality of the whole company."[9] A vice-president of IBM once said that Tom Watson, Sr.'s "personality and force have saturated . . . IBM until now the personality of the man and the personality of the corporation are so closely identified as to be practically one and the same."[10]

WHY CULTURE IS SO IMPORTANT

Culture provides consistency for an organization and its people. Leaders have several paradigms to choose from, but they typically stay with one, particularly when the guiding system that they have chosen is internally consistent and provides evidence of success. The more the organization succeeds, the stronger the commitment to the culture that brings that success.

An organization's culture provides order and structure for activity. It provides people with an internal way of life and, in so doing, plays the same role for people that a society's culture plays. It tells people which activities are in bounds and which are out of bounds. It establishes ground rules for people, determining what is right and what is wrong. Over time, a culture establishes communication patterns—the kind of language people use with one another and the assumptions upon which they consistently operate. It establishes membership criteria, who is included and who is excluded.

A culture establishes the conditions for judging internal effectiveness. It determines whether performance is effective or ineffective, and what effective and ineffective mean in the organization. It sets the expectations and priorities—what's important around here—and conditions for reward and punishment. People who adhere to these expectations and priorities get promoted and advance; those who don't adhere, don't advance. Too much noncompliance typically leads to termination.

Culture determines the nature and use of power within an organization. It fixes power at the top of an organization, disperses power throughout, or some of both. It gives the organization more power than individuals or individuals more power than the organization and installs the process for how decisions get made.

Culture has much to say about the structure of an organization—who reports to whom and the alignment of functions with one another. It directly or indirectly defines status within the organization—what it is and how one acquires it.

It sets the pattern for how people and functions relate to one another. It emphasizes territory or it does not. Culture lets people know how close they can get to one another, and determines whether or not teaming is important and expected. Culture also provides the framework for addressing, managing, and resolving conflicts in the organization.

Culture establishes management practices. It fixes how an organization plans its work, organizes and coordinates activity, manages performance, and gets the results it deems important.

Culture is also important because it limits strategy. Organizations that "stick to their knitting"[11] are operating consistently with their culture and its inherent constraints. Many organizations have learned the hard way that new strategies that make sense from a financial, product, or marketing viewpoint cannot be implemented because they are too far out of line with the organization's existing assumptions. The new strategies do not fit within the cultural paradigm. This certainly has been the case with organizations that have been around awhile and which have been historically successful in achieving their goals.

The failure of 50 percent of the mergers and acquisitions between organizations during the 1980s is testimony to the importance of culture and its limiting role on strategy.[12] A cultural mismatch in an acquisition or merger is perhaps as great a risk as a financial, product, or market mismatch. The merger between Gulf Corporation and Chevron Corporation was initially regarded as a near-perfect match because the two oil giants had approximately equal assets and complementary resources. Later, however, the merger was described as a "forced marriage," laden with fears, anxieties, and frustrations, because the companies approached the same business with widely differing styles and strategies.[13] Virtually identical dynamics were characteristic of the merger that created LTV Steel, the second-largest steelmaker in the United States. While the combination and "all the numbers look[ed] good on paper," Republic Steel and Jones & Laughlin Steel, the merger partners, had sufficiently different philosophies, styles, and orientations so that the simplest disagreements between employee groups often flared into "major conflicts."[14]

The merger between Rockwell International and North American in 1968 is another example.

When marketing-oriented Rockwell International merged with aerospace engineering wizards at North American, managers and analysts alike expected a synergistic reaction . . . Rather than supporting each other, however, the basic values of the firms collided. As then-CEO Robert Anderson lamented, the aerospace people weren't used to commercial problems: "We kept beating them on the head to diversify, but every time they'd try it they'd spend a lot of money on something that . . . there was no market for, or they overdesigned for the market." The world views of the two firms . . . were radically different. Rockwell's company culture looked at the world as a rough-and-tumble place where profit margins dominate decision-making. North American's environment was more noble. Some 60 well-paid PhD's . . . spent only 20 percent of their time on company business and were free to devote the rest as they chose to basic research. This was not compatible with Rockwell's obsession about controlling costs and margins. Thirteen years later, executives are still trying to improve the cultural fit of the two firms.[15]

The management of productivity, which Peter Drucker says is the primary function of management,[16] is fundamentally a cultural phenomenon. Culture determines how success is defined and accomplished, and productivity is tied directly to an organization's formula for success.

Organizational culture parallels individual character. Culture serves an important and fundamental organizing function for an organization just as character does for the individual. The parallel between culture and character is discussed much more fully in Chapter 8.

Appendix to Chapter One
WHY CULTURE IS SO IMPORTANT

It provides consistency for an organization and its people.
It provides order and structure for activity within an organization.
It establishes an internal way of life for people.
    It provides boundaries and ground rules.
    It establishes communications patterns.
    It establishes membership criteria.
It determines the conditions for internal effectiveness.
    It sets the conditions for reward and punishment.
    It sets up expectations and priorities.
    It determines the nature and use of power.
It strongly influences how an organization is structured.
It sets the patterns for internal relationships among people.
It defines effective and ineffective performance.
It fixes an organization's approach to management.

It limits strategy.
It is fundamental to an organization's productivity.
It parallels individual character.

NOTES

1. The term *organizational culture* is used interchangeably with the term *corporate culture* in this book. Both concepts pertain not only to business organizations but to every kind of organization.

2. The first part of this definition coincides exactly with the definition of culture proposed by Marvin Bower, *The Will to Manage* (New York: McGraw-Hill, 1966). Bower's definition, "the way we do things around here," grew out of many years of consulting with a wide variety of organizations.

3. The natural suitability of a core culture for a particular kind of endeavor is an area of investigation that holds considerable promise. See the discussion at the end of Chapter 8.

4. Ed Schein, an authority on organizational culture insists that "leadership and culture are two sides of the same coin." He believes that "there is the real possibility that the only thing of real importance that leaders do is to create and manage culture." See Ed H. Schein, *Organizational Culture and Leadership* (San Francisco: Jossey-Bass, 1985), 2.

5. T. S. Kuhn, *The Structure of Scientific Revolutions*, 2nd ed. (Chicago: University of Chicago Press, 1970), 11.

6. There are precedents for this premise. Max Weber postulated the view that bureaucracy—order by rule—is the most efficient form of human organization. He specifically touted the military as a model for organizations. See Max Weber, *The Theory of Social and Economic Organization* (New York: Oxford University Press, 1947). Henri Fayol also proposed the military as a model for organizing work in *General and Industrial Administration* (New York: Pitman, 1949). In a more recent article, D. J. Garsombke comments on the power of the military model in American business organizations: "there is a militarism cast to organizational culture in the United States . . . [that is] so much in evidence that one must examine its impact both on organizations themselves and on the people who work in them." See D. J. Garsombke, "Organizational Culture Dons the Mantle of Militarism," *Organizational Dynamics*, 17, no. 1 (Summer 1988), 46. Bennis and Nanus suggest the prevalence of a "collegial" culture that is

"typically formed by leaders who bring academic and scientific values to the workplace." See W. Bennis and B. Nanus, *Leaders: The Strategies for Taking Charge* (New York: Harper & Row, 1985), 119–123.

7. D. C. McClelland, *The Achieving Society* (Princeton, NJ: Van Nostrand, 1961); *Power: The Inner Experience* (New York: Irvington, 1975); and D. C. McClelland and D. G. Winters, *Motivating Economic Achievement* (New York: Free Press, 1969); A. H. Maslow, *Motivation and Personality* (New York: Harper, 1954); *Toward A Psychology of Being* (Princeton, NJ: D. Van Nostrand, 1968).

8. Terrence E. Deal and Allan A. Kennedy, *Corporate Cultures: The Rites and Rituals of Corporate Life* (Reading, MA: Addison-Wesley, 1982), 13.

9. Quoted in F. G. Harmon and G. Jacobs, *The Vital Difference: Unleashing the Powers of Sustained Corporate Success* (New York: AMACOM, 1985), 36.

10. Ibid., 36.

11. "Sticking to [one's] knitting" is one of eight qualities characteristic of "excellent" companies. See T. J. Peters and R. H. Waterman, Jr., *In Search of Excellence: Lessons from America's Best-Run Companies* (New York: Harper & Row, 1982).

12. Fifty percent is a conservative figure according to P. Pritchett, who has done a great deal of work on mergers and acquisitions; see *After the Merger: Managing the Shockwaves* (Homewood, IL: Dow Jones-Irwin, 1985). A study of joint ventures formed between 1924 and 1985 found that only 46 percent succeeded. The average life span was only three and one-half years. See K. R. Harrigan, *Strategies for Joint Ventures* (Lexington, MA: Lexington Books, 1985).

13. K. Wells and C. Hymowitz, "Takeover Trauma: Gulf's Managers Find Merger into Chevron Forces Many Changes," *The Wall Street Journal* (December 5, 1984).

14. T. F. O'Boyle and M. Russell, "Troubled Marriage: Steel Giants' Merger Brings Headaches, J&L and Republic Find," *The Wall Street Journal* (November 30, 1984).

15. Quoted in W. Bennis and B. Nanus, *Leaders: The Strategies for Taking Charge* (New York: Harper & Row, 1985), 114–115.

16. Peter F. Drucker, *Managing in Turbulent Times* (New York: Harper & Row, 1980), 14.

Chapter Two

Core Culture:
A Questionnaire

The questionnaire is designed to help you identify the core culture that exists within your own organization quickly and in a useful way. When you complete and score the questionnaire, go directly to the chapter that describes the core culture determined by your score.

This questionnaire is a tool to make the concepts of this book personal and concrete. Early discovery of the core culture that exists within your organization allows you to work your way through the remaining material with your organization in mind. At a minimum, the questionnaire will add the dimension of your personal experience as a continuing reference point for the concepts ahead.

This questionnaire is not a scientifically validated instrument but a helpful beginning and a guide as you read through the balance of this book.[1] The items in the questionnaire are derived directly from the concepts of the four core cultures and have the same value and utility as the concepts themselves.

You may wish to ask others in your organization to complete copies of this questionnaire. The results would serve as a comparative base. On items where your colleagues differ, discuss the nature of these differences. This should help clarify what kind of core culture exists within your organization. The way your colleagues see the behavior of your organization may cause you to change your response to some items in the questionnaire; it could of course work the other way around.

Scoring

Assuming that your organization has been in existence for at least a year and is functioning in a fairly effective manner, you should score at least 50 percent or greater of the items on one of the four core cultures. If this does not happen, it may be not only advised but necessary to ask others to fill out the questionnaire independently in order to identify your organization's core culture. If your score does not yield a clear majority, it is likely that your organization presently has two core cultures. It is unlikely that you will have more than two.

Another approach is to read the chapters on the two core cultures that emerge from your score and then decide which of the two is actually present within your organization.

CORE CULTURE QUESTIONNAIRE

Instructions

This questionnaire is designed to help you determine the core culture that exists in your own organization. As you work through the questions, keep the following in mind:

- "Core culture" is defined as: How we do things in order to succeed.
- For each question ask yourself: "When I boil it down and get to the heart of the matter, which of the four possible answers most accurately describes my actual experience in my organization?" Your organization may have a little of each of the four possible answers, but you need to determine which is the truest for your organization. Your actual experience in the organization is very important, so concentrate on how things are, not on how you believe they ought to be.
- Be objective. Each core culture has its own mix of strengths and weaknesses, and you are trying to determine only the kind that exists in your organization.

- Focus on your whole organization. These questions are
  not about your department, your group, your division, or
  your unit but about the organization as a whole.

Answer every question, and select only one answer for each.
When in doubt, go with the first response you had when you
read the question.

Questionnaire

1. When all is said and done, the way we accomplish success
   in this organization is to:
   _____ a. Get and keep control.
   _____ b. Put a collection of people together, build them
            into a team, and charge them with fully
            utilizing one another as resources.
   _____ c. Create an organization that has the highest
            possible level of competence and capitalize
            on that competence.
   _____ d. Provide the conditions whereby the people
            within the organization can develop and
            make valuable accomplishments.

2. What do we pay attention to primarily in our organization
   and how do we decide about things?
   _____ a. We pay attention to what might be and we
            decide by relying on objective and detached
            analysis.
   _____ b. We pay attention to what is and we decide
            by relying on what evolves from within the
            hearts and minds of our people.
   _____ c. We pay attention to what might be and we
            decide by relying on what evolves from within
            the hearts and minds of our people.
   _____ d. We pay attention to what is and we decide by
            relying on objective and detached analysis.

3. The people with the most power and influence in the
   organization:
   _____ a. Are charismatic, can inspire others, and
            are good at motivating others to develop their
            potential.
   _____ b. Have the title and position that gives them
            the right and the authority to exercise power
            and influence.

_____ c. Are both contributors and team players,
who are an essential part of the team. People
like working with them.

_____ d. Are experts or specialists, who have the most
knowledge about something important.

4. In our organization, "success" means:

_____ a. Synergy. By teaming up with one another and
with our customers, we accomplish what
we are after.

_____ b. Growth. Success means helping others more
fully realize their potential.

_____ c. Dominance. Success means having more
control than anyone else. Complete success
would be for the organization to be the only
game in town.

_____ d. Superiority. Success means that the
organization is the best, offering superior
value. The organization is "state of the art"
in all that it does.

5. In our organization, leadership means:

_____ a. Authority. Leaders are regulators and call
the shots. They are commanding, firm, and
definitive. What they say goes.

_____ b. Setting standards and working hard to get
people to achieve more. Leaders are intense
taskmasters, who always challenge workers
to be better.

_____ c. Being a catalyst. Leaders cultivate people.
They create conditions in which people are
inspired to fulfill their own and others'
potential. At the same time, leaders build
commitment to the organization.

_____ d. Building a team that will work well together.
Leaders are coaches. They behave as
first-among-equals. They strive to represent
the people in the organization.

6. When we worry about something in the organization, it is
usually about:

_____ a. Losing. We worry most about being also-rans
or having our reputation harmed because
we couldn't deliver as well as, or better than,
our competitors.

_____ b. Stagnation. We worry most about failing
to progress, simply existing from day to day,
or even going backwards.

_____ c. Vulnerability. We worry most about being in
a position where others have more power or
market share than we do.

_____ d. Lack of unity. We worry most about the
team being broken up or alienating our
customers. We worry about a lack of trust
among ourselves.

7. Our organization's overall management style is best
described as:

_____ a. Enabling. Empowering. Commitment oriented.

_____ b. Challenging. Goal oriented. Very rational and
analytical.

_____ c. Democratic. Highly relational. Highly
participative.

_____ d. Prescriptive. Methodical. Policy and procedure
oriented.

8. The essential role of the individual employee in our
organization is to:

_____ a. Collaborate. To be a team player.

_____ b. Be an expert. To be the best in your specialty
or area of technical expertise.

_____ c. Perform according to policy and procedure. To
meet the requirements of the job as outlined.

_____ d. Be all you can be. To change, develop, and
grow. To be committed to the organization
and its purposes.

9. What counts most in the organization is:

_____ a. Winning. Being recognized as the best
competitor around.

_____ b. Not losing. Keeping what we've got.

_____ c. Evolving. Realizing greater potential. Fulfilling
commitments.

_____ d. Accomplishing it together. Being able to say
"we did it together."

10. Which of the following best describes how you feel about
working in your organization:

_____ a. This is a caring and "spirited" place. I feel
supported.

_____ b. People are able to count on one another.

_____ c. Things are no nonsense and restrained.

_____ d. Things are rather intense. I feel like I have to be on my toes all the time.

11. What counts most in the organization is:

_____ a. Security.

_____ b. Community.

_____ c. Merit.

_____ d. Fulfillment.

12. Which of the following best describes the primary way decisions are made in the organization?

_____ a. We pay close attention to our concepts and standards. We emphasize the fit between our theoretical goals and the extent to which we achieve them. Our decision-making process centers on how systematically our conceptual goals are achieved.

_____ b. We pay close attention to our values. We emphasize the fit between our values and how close we are to realizing them. Our decision-making process centers on the congruence between our values or purposes and what we have put into practice.

_____ c. We emphasize what the organization needs. Our decision-making process centers on the objectives of the organization and on what we need from each function within the organization.

_____ d. We emphasize tapping into the experiences of one another. Our decision-making process centers on fully using our collective experiences and pushing for a consensus.

13. Overall, life inside our organization is:

_____ a. Spontaneous, interactive, and free and easy.

_____ b. Intellectually competitive, rigorous, and intense.

_____ c. Objective, orderly, and serious.

_____ d. Subjective, dedicated, and purposeful.

14. In general, our attitude toward mistakes is:

_____ a. We tend to minimize the impact of mistakes and do not worry much about them. People who make mistakes should be given another chance.

_____ b. Mistakes are inevitable, but we manage by picking up the pieces and making the necessary corrections before they grow into bigger problems.

_____ c. Mistakes are nearly taboo. We don't like them. A person who makes mistakes is looked down upon.

_____ d. We pay attention to the kind of mistake. If the mistake can be quickly fixed, we go ahead and fix it. If the mistake causes a function to get in trouble or could cause the organization to become vulnerable, we marshal all our resources to fix it as quickly as possible. Mistakes that affect the organization as a whole could get someone in trouble.

15. Concerning control, which of the following is most emphasized?

_____ a. Concepts and ideas. We control everything that is critical toward achieving or preserving our superiority in the marketplace.

_____ b. Everything critical to keeping us working together in the organization and retaining close ties with our customers.

_____ c. Just about everything. Getting and keeping control is central to what the organization is and does.

_____ d. As little as possible. We are put off by the notion of control. We prefer to leave things up to the commitment and good will of our people.

16. The essential nature of work in the organization emphasizes:

_____ a. Functionalists. Individuals stay within their function. Specialties are subordinate to the service of functions.

_____ b. Specialists. Individuals stay in their technical or other specialty. Functions are channeled into the service of specialties.

_____ c. Generalists. Individuals move in and out of numerous functions and specialties.

_____ d. All of the above. Individuals do all three.

17. The people who primarily get promoted in the organization are:

    _____ a. Generalists. They must also be capable people who are easy to work with.

    _____ b. Those who have performed consistently well in their function for many years and have demonstrated that they can seize authority and get things done.

    _____ c. Those who know the most about their area of expertise and have demonstrated their competence.

    _____ d. People who can handle responsibility and who want it. We don't use the word "promotion."

18. The compensation system in the organization is most similar to which of the following?

    _____ a. We emphasize fair and equitable pay for all. We also emphasize the long-term perspective. We plow a lot of money back into the organization to ensure continued growth and success, so personal financial compensation tends to be secondary to other more important matters.

    _____ b. Our compensation is highly individual and incentive oriented. Uniquely capable people who are recognized experts can make a lot of money.

    _____ c. Our compensation system is highly structured. The larger your role and function in the organization, the more money you make.

    _____ d. Our compensation is tied primarily to team effort. If the whole organization does well, we all share in the wealth. If the whole organization does poorly, we all sacrifice.

19. Which of the following best describes our organization's primary approach in dealing with customers or constituents?

    _____ a. Partnership. We team up with our customers or constituents. We want to be able to say "We did it together."

    _____ b. We emphasize uplifting and enriching our customers or constituents. We concentrate on realizing the possibilities and potential of our customers or constituents more fully.

_____ c. We emphasize gaining the greatest market share that we can get. We would like to be the only game in town for our customers or constituents.

_____ d. We emphasize offering superior value to our customers or constituents. We try to provide state-of-the-art goods or services to our customers or constituents.

20. Which phrase best describes our organization?

_____ a. "We believe in what we are doing, we make a commitment, and we realize unlimited potential."

_____ b. "We are the best at what we do."

_____ c. "We are the biggest at what we do."

_____ d. "United we stand, divided we fall."

Scoring Table

Question	Core Culture			
	I	II	III	IV
1	a	b	c	d
2	d	b	a	c
3	b	c	d	a
4	c	a	d	b
5	a	d	b	c
6	c	d	a	b
7	d	c	b	a
8	c	a	b	d
9	b	d	a	c
10	c	b	d	a
11	a	b	c	d
12	c	d	a	b
13	c	a	b	d
14	d	b	c	a
15	c	b	a	d
16	a	c	b	d
17	b	a	c	d
18	c	d	b	a
19	c	a	d	b
20	c	d	b	a
Total				

Directions for Scoring the Questionnaire

Record your answer to each question (a, b, c, or d) on the Scoring Table by writing an "X" across the letter chosen at the proper question number. Do this for every question.

When finished, add up the total number of X's recorded under each roman numeral at the top of the Scoring Table. Roman numeral I relates to the control culture, roman numeral II to the collaboration culture, roman numeral III to the competence culture, and roman numeral IV to the cultivation culture. The roman numeral with the majority of X's is your organization's core culture. A majority is 50 percent or greater, or 10 out of 20 items or greater. If your scores do not reflect a clear majority in one core culture, reread the first part of this chapter.

NOTE

1. This author is currently developing a more extended version of this questionnaire which is intended for use in future organizational analysis and development efforts. This extended version will not only address the question of which core culture exists within any one organization, but will also identify the extent to which various critical organizational subfactors (i.e., selection of new personnel, approach with customers/constituents, management style, and so on) are or are not congruent with one's organizational core culture.

Chapter Three

The Control Culture

The control culture has its socialization base in military organizations. This refers to armed service organizations, not military think tanks or research and development entities. The military prototype for an organizational culture has existed for centuries, but it became widely prevalent during the two world wars. Today the military is a prototype for many different kinds of organizations in business; education, particularly elementary and secondary schools; government; and numerous others.[1]

The individual motivation base for the control culture lies in people's need for power. Power is one of three fundamental human motivations in work.[2] Power enables an individual to have an impact, control, or significantly influence the means of effecting a person, situation, or organization. The leadership of a control organization values dominance most. Control cultures are imbued with power within their organizations and within the environments that they operate. In business organizations, this often translates into gaining maximum market share. In other organizations, it translates into gaining a position where consumers or others have little choice but to rely on them. Control cultures like to be "the only game in town." Kellogg Company, an American institution and a company much-revered by its employees, posts the company's daily market share in the United States in its cereal plant in Battle Creek, Michigan. At Kellogg, "market share is crucial. A one-point swing, for example, means a gain—or loss—of $40 million in annual revenues to Kellogg."[3] In sum, the natural and automatic meaning of success for a control culture is to gain dominance.

Control cultures abhor feeling vulnerable. A control culture is always occupied with heading off vulnerability or overcoming any sign of it. Control cultures put great stock in procedures that track the organization's environment for possible areas of weakness; for example, enemy tracking systems for the military and polls for politicians. They also quickly suppress discontent or any signs of disruption; many elementary and secondary schools systems operate in this manner. Security has great import in a control culture. Much is done to preserve security and the perception of it. Feeling insecure is, after all, feeling vulnerable.

XYZ Communications, Inc. (a pseudonym for an actual company) is an excellent example of a control culture, particularly financial control. This company built a highly sophisticated financial control system that provides its managers with virtually instant information about every aspect of the company's financial performance. The detailed system tracks the overall financial performance of the company and each of its operating units on a weekly, monthly, quarterly, and annual basis. The control system provides comparative information that goes back four years (e.g., ROI per product line, per quarter, per year, per operating unit, etc.). Thousands of hours of management and employee time have gone into its development and implementation. This control system is central to "how things get done around" this company for the organization and its people to succeed. In many ways, this financial control system is *the* management system of XYZ Communications. At any sign of vulnerability, the company immediately begins to cut costs or analyze the poor revenue generation performance of a particular operating unit or product line.

As one might expect, the customers or consumers of a control culture usually feel they must adapt to the control culture organization—a not uncommon experience for many who deal with the US Post Office or other governmental entities. If the fundamental posture of a control culture is to get and maintain control, then its members will behave in a controlling and dominating manner with customers, consumers, or constituents. Monopolies are often control cultures. Suppliers to monopolies tell similar stories about the intensity with which control cultures work at getting and staying in control on every issue, big or small.

Control cultures also strive to have the only product, technology, or service. Kodak Corporation had a $1 billion dollar fine levied against it in late 1991 for attempting to dominate the instant photography market. Kodak and Polaroid had been locked in a 10-year battle over this technology. Copyrights, patents, and trademarks—legal forms of maintaining control—are important in control cultures. Control culture organizations have large staffs to ensure that patents, copyrights, and trademarks are air tight. Colonel (note the military emphasis) Sanders of Kentucky Fried Chicken and his staff labored in the courts for decades to protect his recipe for preparing chicken. KFC's fundamental posture is to succeed by control.

William Wrigley Jr. Company, the Chicago chewing gum manufacturer produces annually more than 14,000 tons of gum.

> [Wrigley] dominates the world of chewing gum more thoroughly than Coca-Cola Co. looms over soft drinks. Wrigley has a 48 percent share of a $2.4 billion US retail market, nearly twice that of Warner-Lambert Co., its biggest competitor. Mr. Wrigley . . . oversees a paternalistic corporate culture and a secretive manufacturing operation. Wrigley's sprawling red-brick compound in southwest Chicago, one of the 13 gum factories it operates throughout the world, is guarded by security cameras and tall barbed-wire gates. Were it not for a small, brass nameplate and the smell of mint hanging in the air, the factory would be a complete mystery in a mixed-use neighborhood of warehouses and modest old houses. Uniformed security guards patrol the compound in Jeeps.[4]

Wrigley has upgraded its manufacturing equipment in the last five years, but "it won't show its new gizmos to people who don't work there. And it is so worried about its secrets that it doesn't get patents for innovations that others might copy. It is best that nobody else knows anything about them at all, Wrigley figures."[5] The Wrigley Company is a clear expression of a control culture. One would expect barbed-wire fences and security guards in a military compound, but here we have the same concern with security and protection for a chewing gum manufacturing plant.

Control cultures prize objectivity. Emotions, subjectivity, and "soft" concepts take everyone's eye off the ball and potentially get the organization in trouble. Empiricism and the systematic examination of externally generated facts are highly valued. Control culture managers are strong skeptics. They want evidence and proof for everything. In many ways, they operate unconsciously as though they are in a life-and-death enterprise, a feeling carried through by the military paradigm. Such a prevailing attitude is derived, perhaps, from emphasizing control to a great extent or for an extended period of time. People in these cultures are often very serious.

Order and predictability, and maintaining stability are important in control cultures. Perpetuation of the status quo is paramount, and change is resisted. The traditional way of doing things is part and parcel of control cultures. Standardization and routinization are valued. Control cultures are usually very structured environments; they attract people who like structure and order. Control cultures often have a legalistic quality. Documentation has great importance because it significantly enhances control and corrective action.

Control cultures hold strongly to what might be termed "systematism," particularly organizational systematism. The organizational system itself is the priority in these cultures. At Procter & Gamble, the system

> is based on a simple set of values, well understood throughout the organization, and a socialization process designed to pass those values on to successive generations of managers. The result of this system is that individual identity is always minimized and sometimes lost. . . . As one manager put it, "Everyone at P&G is like a hand in a bucket of water—when the hand is removed, the water closes in and there is no trace."[6]

An individual who leaves Procter & Gamble for good on a Friday afternoon would find his office totally changed the following Monday. Both walls and doors are movable panels, so a wall is now where a door was. The manager thought P&G was trying to eliminate the memory of the departed employee completely. "According to another manager at P&G, when people leave the company abruptly, everything that they had worked on could be reconstructed from the files in a matter of hours. The *system* (author's emphasis) just picked up where they had left off."[7]

General Electric, like Procter & Gamble, has been praised for its sound management practices. Early in his career as CEO of General Electric, Reginald Jones stated his management philosophy. Jones's tenets illustrate the inherent systematism present in the culture at GE. Each tenet relates essentially to the priority of the organizational system.

- The imperative of liquidity.
- Reinvest most heavily in businesses where the investment can be recovered.
- Manage a tight portfolio by continued pruning.
- Tighter control and reduction of autonomy at all levels.
- Contribution of "added value" to the total by each higher echelon.[8]

It is to be expected that high-risk and heavily capital-intensive enterprises are often control cultures. The more risk for people, money, and security, the greater the pull toward control. Many energy companies, resource companies, and defense manufacturing companies are control cultures. General Dynamics in Washington, D.C., is a control culture, highly preoccupied with security, protecting information, and protecting its technology for manufacturing materials and equipment for the military.

Utility and pragmatism also are important elements to a control culture. Control cultures prize what is useful and what works.

Accumulation tends to prevail in control cultures. These organizations keep piling things up—bigger is better. Typically, when you have more of something, you have more power and an advantage in the pursuit of domination. Take Exxon: number one on the Fortune 500, the largest industrial company in the world, 62 refineries in 33 countries, 14,000 oil wells worldwide, and subsidiaries in 100 countries. Exxon believes that the organization should consist of interchangeable parts and not rely on exceptional personalities, a policy that has clear parallels with the military. Every Monday afternoon Exxon's top management committee meets to review the progress of the company's top 500 managers.[9]

Exxon's management style is highly methodical, careful, and conservative. Objectivity and impersonality are key company norms. An "appraisal review" of every employee's performance is made on a six-page form that covers 21 categories. Employees are compared with and ranked against their peers in the organization. All employees are ranked on a Bell curve; that is, only 10 percent of the company's employees can be rated as "outstanding," etc. The ideal "Exxon person" serves the company throughout his or her career in a variety of staff and line positions, and in a variety of locations, including overseas. At Exxon, the individual clearly works for the good of the overall organization, a characteristic of control cultures.

LEADERSHIP, AUTHORITY, AND DECISION MAKING

Leaders in control cultures typically emphasize a conservative, cautious approach to taking the organization into the future. They place great emphasis on the development of clear policies and procedures and on compliance with them. Rules and regulations are important to these leaders.

Leaders and managers in control cultures exercise *role* power. Authority and the power that accompanies it stem from one's position within the organization. Titles and rank have great importance. Often the size, shape, and furnishings of one's office are tied directly to one's rank—the higher up you are on the organization chart, the bigger your office and the more expensive your furnishings. The more important your title is in a control culture, the more power, prestige, and status you have.

The overt utilization of reward and punishment in leading and managing others prevails. If you perform according to your role requirements, you are rewarded, usually monetarily. If not, you are punished.

Leaders in control cultures are usually authoritative, directive, and often paternalistic. They command respect and compliance from their subordinates. Their behavior connotes that they know what is best for the organization and its people, and that they are solely in charge and will be held accountable for success or failure. They strive to lead with firmness and assertiveness and to portray confidence in what they are doing. Subordinates usually know where their bosses stand on issues and defer to them. Gaining favor with the leadership is very important in control cultures.

Lines of authority are adhered to strictly. A person stays within the chain of command. Going around one's boss to a higher level of authority is taboo. Delegation is usually very tight, clear, and tied to role and specific function. Individuals who have authority delegated to them typically have a very clear idea of the scope and limit of their authority.

In control cultures, leaders and managers tend to be tough-minded, objective, and realistic; they are personally comfortable with taking command and exercising authority. When a decision needs to be made, they can and will make it. They know that their people rely upon them for setting direction and calling the shots. They are perceived as people who can bring order and stability out of chaos and provide the organization with a specific, well-defined sense of direction.

Information within control cultures is guarded with great care, particularly information deemed critical to the success of the organization. You have read that control cultures go to great lengths legally to protect their technologies, manufacturing processes, and other secrets. Secrecy and information protection often spills over into other aspects of control culture life; accordingly, people behave secretly and protectively about most things most of the time. In addition, the flow of information in control cultures is typically downward and upward and less horizontal, so that peers often feel left out of the loop.

The decision-making process in control cultures is thorough, methodical, and systematic. A high value is placed on taking the realistic perspective on issues, and people who are capable realists do well. Decision makers try to have a single focus for making judgments and decisions. Facts count. Decision makers take a long time to gather and systematically analyze the facts. Hip shooting is decried. Solutions that meet current needs are emphasized. Data prevail over theories. Decision makers emphasize the accomplishment of concrete, immediate, and tangible results. What *is* prevails over what *might be*. The objective of decision making is to obtain the single best answer or optimal solution.

Control cultures concentrate on specifics and details. Thoroughness is prized. Typically, as the organization attains more stability, order, and dominance, decision making becomes more thorough and systematic. Certainty and specificity are highly valued. So are past experience and the traditional way of doing things. Hard-headed decision makers would rather test a situation against established principles or talk with someone who has been through a problem before than waste time coming up with a unique solution. They dread surprises. Reasoning is from cause to effect, from premise to conclusion. People who don't have their facts to back up a solution quickly lose respect. Decision making is highly detached and impersonal. Recommending something because it "feels right" is not a good idea in a control culture.

Leaders at the top plan and set goals and objectives in a control culture. They often elicit input from below in the form of reports and detailed analyses. Planning is accorded great importance; it is very often treated as *the* way to get and keep control within and without the organization. Leaders and managers go to great lengths thinking things out beforehand, and they try to anticipate and engineer the desired outcomes and processes for getting there. The less left to chance the better, and no surprises. Bigger and more established organizations will have more long-range planning and pay more attention to holding on to what has already been acquired or accomplished.

Colgate-Palmolive in New York is illustrative of front-end control. Reuben Mark, chairman and CEO,

> runs a small team of executives in New York and at the corporate technology center in Piscataway, N.J., in a manner akin to the old German general staff, making detailed plans while keeping in constant contact with his field marshals . . . The German general staff kept its battle plans in long tubes. Mark keeps his in what is called a "bundle book." In it . . . is everything any Colgate country manager anywhere in the world needs to know about a product. "The bundle," says John Steel [Colgate's global marketing manager], "has the product attributes, the formula for making it, packaging standards, market research and the copy points we need to make in our advertising. When a country manager gets a bundle book, he or she can hit the ground running." . . . While the bundle book lays out the broad strategy, tactical considerations are left to the manager in the field . . . Such attention to detail has made Colgate No. 1 worldwide in the $3 billion toothpaste market, with a commanding 43 percent share, up from about 29 percent only five years ago.[10]

Considerable value is placed on building from strength and staying with what has worked in the past. Goals and objectives are realistic and, more often than not, primarily economic or utilitarian. They also are often oriented toward precise measures of input to output. Plans themselves are often highly structured, specific, and very thorough. A great deal of attention is given to analyzing the competition because competitor behavior is critical to gaining dominance. This is less the case, however, in control cultures that have already achieved monopoly or near monopoly status.

STRUCTURE AND RELATIONSHIPS

Control culture organizations are usually structured in a hierarchical fashion. As they grow, they become prone toward bureaucracy. This is particularly true in government and education, particularly elementary and secondary. Roles and functions are defined and refined with considerable specificity and clarity.

More often than not, this culture breeds functionalists. Expertise is important insofar as it serves functional purposes. As a result, control cultures tend to foster functional specialists who are highly skilled at a particular function. Most people in control cultures stay in one function all their careers. If your superiors ask you to move from one function to another, it is likely that they see you as executive material. What counts in a control culture is that the employee serves in a functional capacity to advance the goals of the organization.

Consider the importance that ABC Pipeline Company (a pseudonym) places on functionalism. This company rarely moves people from one function to another. Most people spend their entire careers within one function, and they know this function inside and out. If they leave, they become hard to replace. ABC Pipeline has about 3,000 employees of which no more than 10 have switched functions in the past 10 years.

Policy and procedure are typically very important within control cultures. Thick and detailed manuals abound. Great emphasis is given to ensuring that all employees know exactly what their job is. Control cultures leave no stone unturned in letting people know about organization rules and regulations.

Marriott Hotels Corporation is another control culture. It has developed elaborate policy and procedure manuals for every aspect of its hotel business. Extremely detailed manuals outline for managers and employees exactly what is expected in each function. Adherence to policy and procedure is closely monitored by upper management. A set of finely tuned systems and standard operating procedures (SOPs) control quality at Marriott. The senior vice president for human resources, Cliff Erhlich, said, "We are the biggest SOP company in the universe. SOPs have been our absolute survival during the period of rapid growth." A hotel manager who had previously worked for another major hotel group indicated that compared with other hotels, Marriott's SOPs manuals were like a mountain.[11] The company even has SOPs for friendliness! Individuals within control cultures adapt to roles and functions and not vice versa.

People relationships within functions are often close and cohesive, but relationships between functions tend to be distant and formal. Control cultures are prone toward territoriality. In large control cultures, functions take on a life of their own and may look and act like miniorganizations. Single individuals between functions will get close to other individuals. But whole functions rarely get close to whole functions. At the top of control cultures, management committees are often created to foster integration between functions. This doesn't work out easily, however, because the perspective of functional managers has become so ingrained. The result is that most key decisions are made by people at the top who are usually the only real integrators in the organization.

As one might imagine, teams are most well developed within functions and many become quite effective. Individual commitment, predictably, is strongest within functions and to the organization as a whole.

The climate in a control culture tends to be serious, subdued, low-key, and matter-of-fact. People do not talk more than necessary, and often an air of secrecy prevails. A control culture *feels* controlled—everything in its proper place; people cordial but quiet and discreet; restraint everywhere.

STAFFING AND PERFORMANCE MANAGEMENT

Recruitment and staffing is usually carefully thought out beforehand; job descriptions are developed and criteria for acceptance are nailed down. Outside recruiting firms are often used, particularly for higher-level positions. A candidate for a position is usually interviewed by people within the function and by bosses farther up in the hierarchy.

Control cultures try to promote from within a function. At Detroit Edison, for example, career tracks for people "are quite clearly defined, and managers often know well beforehand what their next position will be . . . This helps create a high level of stability and predictability, both in the organization and in individual careers."[12] Given the nature of a control culture, this is understandable. If people spend their whole careers within a function, the only way for career advancement—within the organization anyway—is to get promoted within their function. People quickly become demoralized when someone from outside the organization or the function is brought in and placed above them.

Choosing a candidate for a position is principally a matter of whether the individual is right for the potential assigned role. Functional fit is usually the key issue in a control culture. Decision making about the right person for the job is typically thorough and detailed. The people making the selection pay great attention to factual material and to the candidate's past accomplishments.

People who fit easiest into a control culture can be characterized as realistic, down-to-earth, factual and logical, steady, and low-key. They tend to prefer structure and definitiveness. They are usually hard workers who enjoy working with facts and day-to-day realities. They like tangible work: products, numbers, facts, objects, machines, and materials. They also prefer working for stable and predictable organizations. They are often very systematic and thorough. They do their own work in an orderly, organized fashion, at a reasonable, regular pace. Financial security is frequently a key motivator for control culture people. And, more often than not, they are motivated principally by power and control.

Control cultures review performance very thoroughly and methodically. Performance reviews tend to be formal and objective. Evidence must be provided. The key emphasis in performance reviews is how completely and effectively an individual has fulfilled the requirements of his or her role. Performance management in a control culture is oriented toward following through; subordinates in control cultures are tracked closely and system atically. Leaders and managers want to ensure that delegated assignments are carried out properly and completed on a timely basis. Not much gets taken for granted. Corrective action is taken quickly. People are expected to perform in a disciplined fashion and adhere to the organization's policies and procedures.

Electronic Data Systems (EDS) in Dallas is a control culture. It has long had a strong military-like culture, and many of its executives are former military personnel. Employee performance in EDS is heavily structured and controlled. Salesmen and women are all required to sell in the same manner. Sales personnel must memorize certain facts about their prospective customers. Performance is closely monitored and, if ineffective, quickly corrected. Discipline pervades EDS culture.

Conflict in control cultures is often a problem. There is a strong tendency to suppress conflict, quickly dispense with it, or contrive to eliminate its possible occurrence on the front end. At Detroit Edison open conflict "has traditionally not been tolerated, and thus issues that might raise conflict often have been avoided."[13] People in control cultures quickly get the message not to bring up matters of conflict. This can lead to misinformation. For example, in large organizations information can get distorted by the time it reaches top management because people down the line refrain from "telling it like it is." In control cultures, conflict is not something to be managed; it is to be suppressed, dispensed with quickly, or precluded on the front end.

Individual development in control cultures is focused primarily on developing functional expertise and ensuring an understanding of overall organizational goals. Development tends to be formalistic, didactic, and highly structured.

Control cultures manage performance by imparting rewards and sanctions. They get what they want by rewarding people for doing something and punishing them for not doing it. Few boat rockers are found in established control cultures. Subordinates feel compelled to comply and stick to business.

WHERE DOES A CONTROL CULTURE FIT?

The control culture has been the subject of endless bashing for the past 10 to 15 years despite the fact that it has been singularly responsible for most organizational accomplishments worldwide since it was first written about in the Book of Exodus. Some, notably Elliott Jacques,[14] have extolled the inherent strength and value of this culture, but Jacques is more the exception than the rule.

The control culture is just as much a leadership alternative for forming an organization as any of the other three cultures. However, it appears that the control culture has been overused as a paradigm for building and running an organization. This overuse has probably caused it to work ineffectively when forced upon a particular enterprise that more advisedly would have employed one of the other three cultures.

A number of enterprises stand out for which this culture is suited naturally. They include commodity or commodity-like enterprises and enterprises that have to do with matters of life and death (e.g., constructing bridges, medical surgery, etc.). It also seems more appropriate for organizations trying to make their way in mature markets. Despite its tendency toward overuse the reader is asked to give the control culture its proper due.

Appendix A to Chapter Three
STRENGTHS OF THE CONTROL CULTURE

It emphasizes strength and the development of strength itself. When it succeeds, it garners considerable strength and stability.

It is very effective at planning.

It does a very good job at building and implementing systems, policies, and procedures.

Because a control culture is so vigilant, it does well at spotting problems and at taking corrective action. When something isn't working, the culture is poised to fix it and get it back on track.

It is very good at getting proprietary markets, technologies, and processes.

When successful, it does a good job of providing people with short- and long-term financial security.

It is orderly and predictable. People feel safe in a control culture.

Expectations, roles, and jobs are clear.

A control culture emphasizes what works.

It doesn't get fooled much. Surprises are kept to a minimum. All the bases get covered.

Decision making is conservative, thorough, and highly realistic. Decisions are carefully thought through beforehand. It is a systematic culture.

People within a control culture become very proficient at their functions. Functional expertise can reach great heights.

Intrafunction teams generally become quite proficient.

If successful, a control culture gets and keeps control, and more often than not it gains a dominating position in its market.

It is a well-organized culture.

Work and results are closely monitored. Following through works naturally.

Things are usually clear-cut and unambiguous. People know what is expected of them.

It is objective and realistic.

It is very effective in high-risk enterprises.

When kept lean and mean, it is very effective at mobilizing decisive action.

Appendix B to Chapter Three
WEAKNESSES OF THE
CONTROL CULTURE

In excess, it overemphasizes trying to control and dominate. This leads to dysfunctionally competitive behavior.

When things don't go as planned and the control culture becomes excessive, distrust and paranoia escalate.

It is prone to fostering too much compliance and getting distorted information from within when things don't go as planned because people are reluctant to give bosses bad news.

Authoritarian leaders stifle differing judgments about critical issues and get told only what they want to hear.

Innovation is low, particularly for purely technical and human resource management innovation. It is more innovative in functional and financial matters.

A successful and, particularly, large control culture is prone toward treating people in an arrogant and cavalier manner. It gives people, especially outsiders, the message that "we can take you or leave you."

It is difficult to be a generalist. It takes years and entails promotion to the highest levels of the organization.

It is impersonal. There is little family feeling. Feelings, subjective ideas, and intuitions are decried.

It is hard to disagree and to voice conflict. The message is that "conflict and disagreement don't cut it."

In excess, it doesn't foster people helping one another out. An excessive control culture fosters a Darwinian norm where survival of the fittest is paramount, and asking for and giving help is taboo.

Excessive control cultures dictate to or ignore customers.

In excess, it takes the fun out of working.

It ignores possibilities. What might be gets lost.

In excess, it becomes inflexible and rigid.

In excess, preservation of the status quo becomes paramount. Ideas for changing something are quickly rejected.

If too intense, people easily feel manipulated or coerced.

In excess, people at lower levels will delegate responsibility upward and refrain from taking responsibility for the results.

Information flow between functions is overly restricted.

Good ideas from below get lost.

It is prone toward bureaucratization.

Individualists have a hard time of it.

In excess, it is caught up in overemphasizing internal organizational issues and needs. It becomes unbalanced by neglecting external environmental issues (e.g., competitors, marketplace, etc.).

NOTES

1. See especially Max Weber's *The Theory of Social and Economic Organization* (New York: Oxford University Press, 1947). Weber is widely regarded as the father of the "military metaphor" for structuring organizations. He placed considerable emphasis on coordination and control, and believed organizations had a better chance to succeed if they reduced diversity and ambiguity.

2. See D. C. McClelland, *Power: The Inner Experience* (New York: Irvington, 1975). Readers familiar with McClelland's work may find that this book does an injustice to his research into the four stages of the

power motive. Nevertheless, the four stages are concerned with the concept of having impact on others, on a marketplace, or on one's subordinates. The author relies on McClelland's definition to capture the importance of the power motive for the control culture.

3. R. Levering and M. Moskowitz, *The 100 Best Companies to Work for in America* (New York: Currency/Doubleday, 1993), 220.

4. B. Pulley, "Impulse Item: Wrigley Is Thriving, Despite the Recession, in a Resilient Business." *The Wall Street Journal* (May 29, 1991), sec. A.

5. Ibid.

6. Quoted in D. R. Denison, *Corporate Culture and Organizational Effectiveness* (New York: John Wiley & Sons, 1990), 151.

7. Ibid.

8. H. Levinson and S. Rosenthal, *CEO: Corporate Leadership in Action* (New York: Basic Books, 1984), 33.

9. R. Levering, M. Moskowitz, and M. Katz, *The 100 Best Companies to Work for in America* (Reading, MA: Addison-Wesley, 1984), pp. 108–11.

10. S. Kindel, "The Bundle Book: At Reuben Mark's Colgate, Attention to Small Details Creates Large Profits," *Financial World* (January 5, 1993), pp. 34–35.

11. F. G. Harmon and G. Jacobs, *The Vital Difference: Unleashing the Powers of Sustained Corporate Success* (New York: AMACOM, 1985), pp. 82–83.

12. D. R. Denison, *Corporate Culture and Organizational Effectiveness* (New York: John Wiley & Sons, 1990), p. 139.

13. Ibid., p. 137.

14. Elliott Jacques, *Requisite Organization: The CEO's Guide to Creative Structure and Leadership* (Arlington, VA: Cason Hall, 1989).

Chapter Four

The Collaboration Culture

What do Delta Airlines, Dana Corporation, and Trammell Crow have in common? How could these companies in such markedly different industries have a common quality that makes them similar to one another? The answer is that each is, in its unique way, a collaboration culture.

In 1982, three stewardesses for Delta Airlines announced that they and other Delta employees were pledging nearly $1,000 each to buy a $30 million Boeing 767 jet for the airline. "We just wanted to say thanks for the way Delta has treated us," one of the women explained. By December they had raised enough pledges to buy the 767. Seven thousand employees turned out at the Atlanta airport for the christening of *The Spirit of Delta*.[1] A large number of Delta employees genuinely love their company; they talk continually about the "Delta family feeling."

The collaboration culture springs from the family. We all use the family prototype as a framework for organizing activity. Many organizations in our modern world rely on this prototype as their fundamental vehicle for culture formation. The family serves as a base for a wide range of organizations, profit and nonprofit, public and private. It is particularly prevalent in service organizations, but many family-owned and -operated businesses are collaboration culture organizations.

The collaboration culture is also strongly influenced by another basic prototype—the sports framework. Athletic endeavor has a powerful effect upon how we think about organizing human activity and how we should accomplish desired objectives. This is particularly true with teams and teaming. Many leaders turn to the central notion of the team when they build their paradigms for organizational culture. In most sports, success is reached by building, developing, and making use of an effective team.

Dana Corporation makes axles, transmissions, clutches, frames, engine parts, and universal joints for cars and trucks, as well as a variety of valves, pumps, and motors for industrial equipment. Dana plants have a company store, or company "identity center," that sells dozens of items bearing the Dana logo: baseball caps, T-shirts, running sweatsuits, shoehorns, golf balls, and bumper stickers. The Dana identity centers sell about one-quarter-million dollars' worth of goods a year. The idea behind this identification program is to get Dana people to consider themselves part of a team, and to understand that their real adversaries are their competitors, Eaton or Rockwell, and not people from within the company. Dana president Stan Gustafson says: "The only difference between us and our competition is our people. If our salesmen have more energy, our engineers are more creative, our people more productive, we are going to beat them. It's our team versus their team."[2] Teams of workers and supervisors at Dana meet regularly to discuss ideas about improving productivity and profitability.

D. C. McClelland's research on individual motivation has yielded a second essential motive, the need for affiliation. The affiliation motive is "the motive to establish, maintain, or restore a positive affective relationship with others. It is a strong need to have friends, to have positive, close relationships with other people."[3] This is the individual motivation base at the heart of the collaboration culture. Its parallel with the family and sports socialization bases is easily seen. The collaboration culture's way to success is to put a collection of people together, to build these people into a team, to engender their positive affective relationship with one another, and to charge them with fully utilizing one another as resources.

The natural and automatic meaning of success for a collaboration culture is to gain synergy. *Synergy* means the simultaneous action of separate agencies, a combined or cooperative action or force—to work together. Together, they have a greater total effect than the sum of their individual effects. Simply expressed, synergy occurs when $2 + 2 = 5$. In other words, you get more when people collaborate than when you add only their hearts and minds. The yield is 5 instead of 4. When people care about one another, team up, and put their best thinking and heightened energy together, the result is a kind of chemical reaction and the creation of a whole new compound. Heating a mixture of aluminum powder and metal oxide to the ignition point produces a self-sustaining source of brilliant light and intense heat that cannot be put out by ordinary means. The mixture will burn underwater or in any other environment that would extinguish an ordinary flame. Like this extraordinary flame, the collaborative group does not depend on its surroundings for support, and a greater total effect takes place than the simple addition of individuals to the group. If synergy is present the group becomes a team. A group is only a collection of individuals independently allied with one another; a team is collaborative and synergistic. When the team works, the results go far beyond what a group can accomplish.

The collaboration culture is a natural at building and utilizing diversity, an increasingly prevalent issue in the 1990s. Its success occurs largely because the organization brings in and relies upon people with diverse backgrounds and capabilities. Diversity is a necessary but not totally necessary element to synergy. The team succeeds by combining and synergizing its diversity. A good baseball team, for example, has a strong and diverse mix of talent—pitchers who can start and relieve, hitters who can hit the long ball and hit singles, players who can steal bases, and fielders who can play different positions effectively. But a winning baseball team rarely succeeds by having all singles hitters in its lineup; it is the mix of talent and the effective blending of that mix that fosters success.

CRS Sirrine, Inc., in Houston, Texas, the architecture and construction management firm, is emphatic about teams and diversity. Its professionals work in teams of six to eight people, with each team working on three to six different projects at a time. Members of a CRS team usually see the project through to the end. Bill Peel, vice president of development, sketches three intersecting circles to show how the firm works. The three circles represent designers, technologists, and managers. According to Peel, "Few people are all three; most are one or a combination of two of them. That's why teams are important."[4]

Another consistent characteristic of a collaboration culture is commitment to the whole organization. What counts is dedication to the good of the whole enterprise—"united we stand, divided we fall." People put their hearts and souls into ensuring that the organization accomplishes its desired objectives. Temporary individual advantage is sacrificed for the good of the whole. Selfishness or individual grandstanding will get you in hot water in a collaboration culture.

Commitment also flows from the organization to the individual. Collaboration culture organizations have a deep commitment to their employees and will do all they can to hold onto them. Many have no-layoff policies. In 1982 the airline industry was beset with hard times and many airline companies were laying off thousands of people, cutting pay, and freezing wages. Braniff International filed for bankruptcy. But Delta Airlines stuck to its long-standing policy of no layoffs and declared an across-the-board pay hike of 8.5 percent. Earlier, during the 1973 energy crisis, when Delta's schedules were sharply curtailed, the company reassigned 600 pilots and stewardesses to work at loading cargo, cleaning airplanes, selling tickets, and making reservations.[5] In 1992 and 1993, Delta reversed its no-layoff policy and began to let some of its people go, culminating in the layoff of 600 pilots in March 1993. Like most airlines in this period, Delta was beset with serious financial losses. The layoffs, however, were a last resort and the number of Delta employees laid off was much lower than at other US airlines. Despite the ups and downs in the airline business, Delta and its employees take these moves in stride, and the team commitment remains strong.

Harmony is highly valued in a collaboration culture. People are expected to work hard at channeling their unique talents into what others need and what is best for the entire organization. This culture is like a symphony orchestra, where individual members constantly focus on tailoring their effort to the work of other members and to the requirements of the musical piece.

A collaboration culture bristles at disharmony and "prima donnaism." Destructive infighting and excessive self-interest run counter to this culture's formula for success and method of operating. Such behavior is not tolerated and quickly leads to exclusion.

The collaboration culture works with its customers principally by forming a partnership with them. Like the other three core cultures, it extends its internal way of operating outward to customers or constituents. In the collaboration culture, this means partnering and teaming up. The collaboration culture strives to bring its customers or constituents into the family or the team. Walt Disney Productions is a particularly good example of this process. Disney employees are trained extensively to treat customers as members of the Disney family. They give considerable attention to building harmony and teaming up with customers; visitors at Disney World in Orlando, Florida, learn firsthand how vigorously Disney's friendly employees work to get them involved and actively participating in the live programs. Indeed, the whole business focus of Disney is family entertainment—partnering up with its consumer public of families. It strives to enhance family life by combining the interests of Walt Disney Productions with those of its customers as families. Disney has become so well known for its ability to do this that many other companies visit Disney Productions to learn its "secrets." The only real secret is that Disney is a collaboration culture, relying on the application of its natural internal way of operating to its customers.

Will Raap is the founder and president of Gardener's Supply Company in Burlington, Vermont. Raap believes that Gardener's Supply fulfills its mission by partnering with its customers. It does so by selling tools and providing garden information and advice. The company has test gardens for trying out new gardening equipment, and it wants to expand them and involve more local gardeners.[6]

Northwestern Mutual Life Insurance Company provides a remarkable example of the extent to which a company forms a partnership with its customers:

> When a man died of gunshot wounds, the coroner's investigation found that it was a case of suicide. The man had taken out a Northwestern Mutual life insurance policy, which did not provide coverage in the event of suicidal death. But Northwestern Mutual was not quite satisfied by the coroner's report and decided to launch an investigation of its own. The company concluded that there was a reasonable doubt about whether it was a case of suicide and paid the full value of the policy to the deceased man's family.[7]

Collaboration cultures put a great deal of effort and attention into understanding their customers and constituents better, developing fuller relationships with them, and even building common missions with them. In the 1990s, collaboration cultures are experimenting with some interesting collaborative evolutions such as (1) inviting customers to collaborate on strategic, design, and marketing issues, and (2) networking with a diverse group of manufacturers, suppliers, distribution companies, and sales organizations instead of having all functions housed within just one company.

Before his resignation in 1993, New York City Police Commissioner Lee P. Brown dedicated himself to changing the New York City Police Department from a control culture to a collaboration culture. "I look at the police mission in a context broader than what the textbooks say about protecting life and property by making arrests," said Brown. "The police are a service organization for the city that doesn't close at 5:00 P.M. We have 24-hour-a-day, 7-day-a-week service. So my mission is to use the resources of the department to improve the quality of life of the citizens of this city. To me that means community policing. It means a cop back on the beat, the way it used to be." A critical component in accomplishing this mission was to team up with the citizens of New York. "Just as companies are finding new ways for customers to participate in improvement, we have a virtually untapped resource of community groups, the private sector, and other city agencies, all of which can help us do community problem solving."[8]

Collaboration cultures are particularly common in service organizations such as health care organizations, especially hospitals. For most health care needs, treatment teams are the rule where different medical and nursing functions work together to diagnose and treat problems. No single discipline or function can do it alone: in the emergency room in a hospital trauma center you will see three-, four-, and five-person teams, each person representing different functions. Team work in an emergency room can make the difference between life and death.

At Greater Southeast Community Hospital in Washington, D.C., most of the work done is by teams. Tom Chapman, CEO of the Greater Southeast Health Care System, which owns the hospital, describes the hospital's culture:

[We] work in terms that are focused on the patient. For example, each elderly patient is treated by a geriatric team that includes a doctor, a nurse, a social worker, a dietician, and a physical therapist. In effect, the patient picks the team leader. If the patient's most critical needs are emotional, then the social worker leads the team—not the doctor. That, of course, turns the traditional hierarchy of a hospital upside down. It also allows for an integrated approach to health care. . . . What makes us unique at Greater Southeast is a shared mind-set that says working together we can solve these problems, whether it is the problem of one patient or the whole community.[9]

People interaction is key in the collaboration culture. The culture succeeds by tapping the talent and dedication of its people and by combining that talent and dedication in a team effort. As a result, how people feel about things is important. Leaders pay considerable attention to creating an internal environment where people feel secure and free to connect and be open with one another. Spontaneous behavior and events are tolerated, particularly those that serve the common good.

A collaboration culture is more egalitarian than other cultures. People hold back from teaming up with people they believe are unequal. Status and rank take a back seat. Ideas and concepts are important, but it is people who are nurtured. The belief is that if you nurture people and encourage interaction, ideas and concepts will take care of themselves.

A natural outgrowth of this emphasis is an ongoing focus on people and reality. Less preoccupied with idealistic matters, this culture stays focused on the daily, detailed, real concerns of its people, particularly those that pertain to their working effectively together and with customers and constituents. People take precedence over roles. Collective contribution to common objectives is more critical than adherence to a particular function or staying within a particular department.

The collaboration culture emphasizes building trust. Politics, infighting, and betrayal erode collaboration and take the organization backward. Leaders work hard to earn their peoples' trust and to create conditions in which trust can flourish. Loyalty is prized. Apogee Enterprises in Minneapolis, Minnesota, is a collaboration culture. CEO Don Goldfus describes what is most important at Apogee:

The key word around here is trust. We subscribe to the Pygmalion Theory, or high-expectations management. We attract and hire people who want to be in business for themselves and put the financial strength of Apogee behind them. I can't possibly run all of these companies. Why, I don't know if I could run any of them. So you need people out there who will run them. That's the power of the system.[10]

Pragmatism flourishes in a collaboration culture. There is usually lots of eclecticism and variety. "Whatever works" is emphasized, particularly what *we* decide will work. This culture is good at rolling with the punches. It sees the need for quick change and changing in midstream. Compelling and intense thinking by fully functioning teams enables changes to be made quickly. Initiative is usually high. Individuals or teams will jump in, take the ball, and start running with it. There is often lots of lively interplay, brainstorming, and experimentation. Some things get tried just to see what will happen. This culture is more tactical than strategic. Data, ideas, and theories are used so long as they get people where they want to go; the emphasis is on payoff. Many sales organizations are collaboration cultures because their procedures fit a pragmatic, quick-moving, team-oriented environment.

Southwest Airlines is an example of pragmatism, experimentation, and reliance on team thinking.

At most Southwest ticket counters, there are now boxy, nondescript automatic ticket machines that will take a credit card and dispense a ticket in just 20 seconds. These marvels were not produced by IBM, working under a two-year, multimillion-dollar contract. Southwest employees built them in their spare time, using off-the-shelf computer parts. "The machine was thought up by a bunch of our guys in a bar one night in Denver," says Audy Donelson, Southwest's station manager at Dallas's Love Field. A similar bootstrap effort is now under way to build PCs at corporate headquarters.[11]

LEADERSHIP, AUTHORITY AND DECISION MAKING

Leaders in collaboration cultures place considerable emphasis on team building, working hard to bring in a mix of talent and fostering cooperation and coordination. Integration is their primary focus, as you would expect in a collaboration culture. The multiple focus takes up much of the time of the leadership.

Leaders foster respect for individual differences. They believe that every perspective has merit and value, particularly when integrated with differing perspectives. Building and preserving trust within the organization, a prerequisite to effective teaming and collaboration, is the particular province of the leadership.

Leaders also devote a great deal of attention to building commitment, identity with the organization, and a strong feeling of psychological ownership in the organization. When they succeed, employees feel a great sense of pride in the organization and feel special because they are associated with it. Like the cultivation culture, individual identity is tied to organizational identity in a collaboration culture, much more so than in the control and competence cultures. This is predictable, given its familial underpinnings.

Trammell Crow, founder of The Trammell Crow Company in Dallas, shares a large bullpen area with partners, secretaries, and accountants. According to Crow, "having everybody together is the best way to communicate. Everybody can see everybody else. You're available. You can get more done. A major component of a successful business is love. How are these people going to know I love them if they cannot see me?" Don Williams, managing partner, said, "we're concerned about the whole person. We try seriously to be a business family. We want to have fun. We want to enjoy it. It's our company." At Trammell Crow, people feel special and are treated as if they are special. The company may well have more millionaires per capita than any other in the United States, over 5 percent of its work force.[12]

Managers in a collaboration culture tend to be "company people" who are willing to make personal sacrifices for the good of the larger organization. Middle managers, particularly, dedicate considerable time and attention to linking people together to ensure that the work is coordinated.

While role power is key in a control culture, relationship power is key in a collaboration culture. Individuals and groups gain power and viability by building relationships that continually draw upon the capabilities of other individuals and groups. You will have relationship power if you are a clear contributor, an effective collaborator, and people like and care about you. True authority and the ability to make things happen is tied to the degree of relationship power that one possesses. The people with authority in a collaboration culture are well-liked team players who make an added-value contribution to the whole organization.

Judgment making and decision making are highly participative and collegial. Brainstorming and give-and-take are common. People want to know what others think and say about issues, problems, strategies, and tactics. The collaboration culture is the most democratic of the four cultures. Leaders usually pay a great deal of attention to the ideas and opinions of others; higher-quality decisions will be made when more people are involved and committed. In addition, people are more committed to implementing decisions when they have been asked to play a role in arriving at them.

Motorola, another collaboration culture organization in the Chicago area, is a firm believer in participative management. Despite its size, Motorola has designed and implemented an elaborate process that guarantees everyone's participation.

Motorola employees have been grouped into teams of some 50 to 250 workers. Each employee shares in a common bonus pool with his or her other team members. The idea is that the people in each pool will be responsible for their own performance—as measured by the production costs and materials use that are controllable by the team, by quality, by production levels, by inventory of stock and finished goods, by housekeeping standards, and by safety records. Whenever an idea proposed by a team leads to a cost reduction, or to production that exceeds a target, all team members share in the gains through bonuses that can amount to 41 percent of base salary (the average varies between 8 and 12 percent).[13]

Motorola has made participation the cornerstone of its culture.

Rod Canion, former CEO of Compaq Computer Corporation, insists that teamwork and consensus lie at the heart of Compaq's success:

Compaq stresses discipline, balance, continuity, and consensus. That's the way to survive in an industry that changes as fast as ours does. There are lots of values behind these characteristics, but perhaps the most important is teamwork . . . Our management process is based on the concept of consensus management. The real benefit of the process is not that you get the answer but all the things you go through to get the answer. You get a lot of facts, you get a lot of people thinking, and the result is that everybody owns the decision when you get through. Originally, we used consensus management at the top to address the really tough, critical, long-term decisions. But as people participated in the process, they could see how to use it at all levels. Today it permeates the company all the way down to the manufacturing floor. When something isn't working right, the teams get together and try to figure it out.[14]

Leaders rarely give orders in the collaboration culture. They see themselves as the representatives of their people and organization, and while they still have ultimate accountability, they want to involve everyone in the decision making. The energy generated by participation goes a long way toward helping the organization accomplish its objectives. It is assumed that people want to do their very best and will do so if given the chance. Because leaders believe that ideas are where you find them, the smart thing to do is to look for them. No one believes that all knowledge resides solely with management. Consider the perspective of Rod Canion: "At Compaq, the consensus process does not assume that because I'm the boss, I have the final answer. It's built around a team, and any time there's a team of people, you expect everybody to contribute in one way or another."[15]

Leaders are often simultaneously supportive and tough-minded by striking a balance between encouragement and seeking clarification of employee contributions. They set people-oriented, down-to-earth, and realistic goals.

Given the value of egalitarianism, leaders are inclined to behave as first among equals. Their job is to get people to work together to accomplish organizational objectives.

STRUCTURE AND RELATIONSHIPS

No matter how a diagram of a collaboration culture might look on paper, the real structure is cluster-like. The top layer of management may have a title like "The Office Of. . . ." The culture is difficult to depict because it is what happens between and among the boxes on the organizational chart that counts.

Collaboration cultures tend to breed generalists because contributing to the whole is important. Generalists offer a wide range of skills and can contribute in a number of different ways. United Parcel Service of America is a good example.

> Chief Executive John W. Rogers and his management committee began their careers as clerks, drivers, and management trainees. Instead of becoming narrow specialists, they learned a little of everything from distribution to marketing as they criss-crossed their way up the corporate ladder. The Chief Executive, a 32-year UPS veteran, still does his own photocopying, eats lunch in the cafeteria alongside packagers and junior managers, and shares a secretary. "When decisions have to be made, we get everyone's opinion, and the company feels like a family to a lot of us." says John Tranfo, a staff Vice President who will soon be celebrating his 40th year at UPS. "In management, we hardly have any turnover," he adds.[16]

Gary Gilsdorf, an 18-year veteran, is a typical Dana Corporation manager. He started as a personnel clerk, became a buyer in the purchasing department, then a production line foreman, and in 1984 took a management position in the industrial relations department at corporate headquarters. He says, "You don't have to go through 12 layers of bureaucracy to do everything because of the policy sheet." He adds that Dana makes it easy for a person who wants to have a variety of experiences in the company. "Dana isn't just interested in specialists. There are lots of opportunities for growth. You can be more rounded."[17]

Once a collaboration culture brings in a multiplicity of talent, it keeps building on that talent by ensuring that people know what it takes to make every function in the organization work. The person who can make the most added value contribution is someone who not only has a set of individual talents but also knows how all the parts of the organization work. Therefore, people move around a great deal, particularly at lower levels in the organization. Persistent individual contribution to the team begins to lessen as people move into senior management.

This is a many-hands-make-light-work world. Cohesion and cooperation make winners of everyone. People quickly get involved in their work and with one another. Individuals readily go the extra mile with little hesitation. Help is easy to get and lots of pinch-hitting goes on. People are always ready to help someone solve a problem.

Traditional staff functions are minimal in a collaboration culture. Those that do take place tend to be expertise oriented. Everyone is an expert in his or her own way.

This is a *we* culture, with a great deal of esprit de corps and camaraderie. People have an easy time celebrating together. They feel a special sense of acceptance and pride through their association with one another and with the organization. People are on a first-name basis. They are sensitive to the feelings of others. Close personal ties are often established that persist after people have left the organization or retired.

The climate in a collaboration culture is usually open and direct. Subordinates feel free to challenge the thinking of top management and their colleagues. Individual differences are carefully preserved.

STAFFING AND PERFORMANCE MANAGEMENT

As in a family, people in a collaboration culture often start out young and grow up with the organization. It is not uncommon for an employee to begin as a teenager or out of college and spend 35 to 40 years with one organization. They start out at the bottom and move up.

This culture also emphasizes recruiting and hiring a wide variety of capable people who are either versatile or high-impact but singular contributors. Given the emphasis on generalists, versatility is particularly prized. The more one can do, the better. The recruiting process itself is usually extensive because it is so important to hire someone who will fit in. Delta Airlines takes the task of hiring very seriously.

> For a new class of 50 flight attendants, the company interviews about 500 applicants. Even temporary hires are sent to Atlanta from elsewhere in the country for a final round of interviews before being selected . . . What is Delta looking for? . . . Delta is looking for members of the family. They are looking for people who belong . . . "We're looking for someone that has wholesomeness," says Marvin

Johnson, Assistant Vice President for Employment. "They don't have to be the richest, nor the best educated, because we hire them based on how they fit in and how they will grow with Delta."[18]

Prima donnas or negative politickers don't last long in a collaboration culture. Goldman, Sachs & Company's co-chairman, John Whitehead, says: "We try to make [negative politicking] unattractive. We reward the guys well for working within groups. Those who work well within those activities achieve favor. The financial incentive is to do it together. We try to recruit people who will be good team players. We're not looking for erratic superstars." John Weinberg, the other co-chairman, adds: "We don't like big egos here. We have superstar talent, but they are not out to publicize themselves."[19] If you fit in, work effectively for the good of the whole organization, and develop a sound, workable knowledge of what makes the organization succeed, you move up. If you like quick upward mobility and notoriety, the collaboration culture is not for you. Disloyalty or egocentricity are looked upon as major flaws.

"Can do" people fit in well. So do people who are empathic and sensitive to others and who naturally express appreciation. Interpersonal tolerance is important. Obviously, team players belong. It also helps to be a good communicator.

Conflict and difference occur naturally and automatically, given the fact that diversity and complementarity are so crucial to the organization's formula for success. Diversity is built in on the front end, and considerable emphasis is placed on the effective management and effective integration of these differences. The collaboration culture is a natural at honoring diversity, which has evolved into an important issue in the 1990s. The culture is usually very effective at managing conflict because this is a skill critical to getting the most from your team.

The development of human resources is emphasized, particularly as it relates to the common good and the accomplishment of organizational goals and objectives.

More often than not collaboration cultures are share-the-wealth cultures. Probably many employee-owned organizations belong to this type of culture. Equity sharing and profit sharing tend to be noticeably generous and pay across levels is usually more evenly divided than in other cultures. Huge imbalances between top management and employees would be inconsistent with the nature of the culture. Rewards are tied principally to organizational performance and less so to individual performance. Individuals do well financially when the whole organization does well. If the organization fails, financial incentive rewards are minimal, even for strong individual efforts.

At Reflexite Corporation, a company based in New Britain, Connecticut, employees own 59 percent of the company's stock. Once a month, employee shareholders receive an "owner's bonus", which is 3 percent of the company's operating profit divided by the total shares. The bonus can add several hundred dollars to a worker's income, during a good month, but when earnings are off they get nothing at all. Shareholders also get annual dividends that amount to 20 percent of pretax earnings. Reflexite typifies how a collaboration culture company pulls together in hard times. Because of cutbacks in state highway construction and maintenance during the recession, Reflexite's sales dropped well below expectations and profits were threatened.

> [The company] came up with an extraordinary layoff-averting strategy. In August [1991] the CEO announced a voluntary leave-of-absence plan. With the permission of their supervisors, employees could ask for up to a month (later two months) off, unpaid. Employees on leave would continue to receive full benefits and would maintain their seniority and owner's bonus rights. In September and October [CEO] Ursprung phased in a salary-and-hours reduction plan. For the rest of the year top management would take a 10 percent pay cut . . . Middle managers and other salaried people would take cuts of 5 percent to 7 percent, and the plant would be closed one weekday a month. As part of both programs, Ursprung detailed other ways the company would cut both manufacturing costs and selling, general, and administrative expenses . . . By late fall Reflexite managers were detecting signs of a recovery: orders were picking up, and the backlog was once again building . . . The company closed the first half of fiscal 1992 with profitability intact.[20]

Individual managers and fellow team members review individual performance. Often peer reviews alone are conducted. Performance reviews focus on the extent to which individuals make an added-value contribution to the company's overall goals and objectives and how effectively they team up with others. "Teamwork and cooperation are near to the heart of Northwestern Mutual [Life Insurance's] personality, and the company recognizes that interpersonal skills are essential for fostering these values. At Northwestern Mutual . . . the most common reason for a manager's not getting along is lack of skill in working as part of a team rather than lack of technical competence."[21]

WHERE DOES A COLLABORATION CULTURE FIT?

The collaboration culture is more naturally suited to enterprises that have incremental relationships with their customers. The relationship between the organization and the customer is principally characterized by incrementalism: the two parties come together in an ongoing, one-step-at-a-time, process.

Enterprises that fit logically into the collaboration cultural framework are characterized by immediacy and by the necessity of co-involvement for things to work. Some examples of this are nursing, entertainment, and many personal service enterprises.

Appendix A to Chapter Four
STRENGTHS OF THE COLLABORATION CULTURE

It is naturally effective at building in and managing diversity and conflict.
Dedication is high. Individuals easily develop a dedication to the success of the whole enterprise, a dedication that the organization returns to the individual.
Communication is open, free, and direct.
It is naturally effective at building, developing, and utilizing teams.
Cohesion and coordination are prevalent. Work relationships are harmonious.
People treat one another in a sensitive and caring manner.

When effective, synergy is accomplished internally and with customers. People resources are fully realized.

People work together and build on one another's skills and capabilities.

People typically help one another out.

When successful, it does a good job of partnering with its customers.

It is good at forming alliances with other organizations.

It is usually egalitarian and democratic in nature.

When successful, trust is prevalent.

It is versatile and adaptive.

It listens to people.

Participative management thrives.

Employees often have a strong sense of ownership and personal pride in the organization.

It fosters individual talent and generalist capabilities.

Tasks and functions are well integrated.

Conflict and differences are typically fostered and are well managed.

Monetary rewards are often very generous.

Appendix B to Chapter Four
WEAKNESSES OF THE COLLABORATION CULTURE

In excess, it leads to people getting caught up with being friends with one another. People refrain from holding one another accountable. Performance slips. The culture gets too supportive and is prone to becoming inbred and cliquish.

It is prone toward the short-term. People want to get to the payoff and start pushing too quickly.

In excess, it becomes overcompromising. It slips into making everybody happy.

In the extreme, it is prone to laissez faire management.

If careless, it fails to recognize individual achievement and inadvertently fosters mediocrity. It is prone toward stifling individuality. An outstanding individual performer loses motivation. Highly talented individualists leave.

It inclines toward de-emphasizing planning.

It gravitates toward groupthink. People refrain from dissent for fear of team ostracism.

In times of difficulty or stress, it gets hamstrung and has a hard time making and sticking to firm decisions.

It is at a disadvantage when competing with a ruthless adversary.

In excess, it takes a long time to make decisions.

Without a central focus, it gets overcommitted and goes off in too many directions.

In excess, it is overadaptive and lets the environment sway and influence it.

NOTES

1. R. Levering, M. Moskowitz, and M. Katz, *The 100 Best Companies to Work for in America* (Reading, MA: Addison-Wesley, 1984), p. 74.
2. Ibid., p. 65.
3. T. R. Horton, *What Works for Me: 16 CEOs Talk about Their Careers and Commitments* (New York: Random House, 1986), p. 405.
4. Levering, Moskowitz, and Katz (1984), pp. 58–59.
5. Ibid., p. 75.
6. U. Gupta, "Keeping the Faith," *The Wall Street Journal Reports* (November 22, 1991), p. R16.
7. F. G. Harmon and G. Jacobs, (1985). *The Vital Difference: Unleashing the Powers of Sustained Corporate Success* (New York: AMACOM, 1985), p. 49.
8. A. M. Webber, "Crime and Management: An Interview with NYC Police Commissioner Lee P. Brown," *Harvard Business Review* (May–June 1991), p. 117. We might speculate that Brown's resignation occurred because his attempt to inculcate a collaboration culture within the Police Department met with considerable resistance at all levels.
9. N. A. Nichols, "Profits with A Purpose: An Interview with Tom Chapman," *Harvard Business Review* (November–December 1992), p. 94.
10. R. Levering and M. Moskowitz, *The 100 Best Companies to Work for in America* (New York: Currency/Doubleday, 1993), pp. 21–22.
11. E. O. Welles, "Captain Marvel," *Inc.* (January 1992), p. 47. Students of culture and organizations will see that the collaboration culture is very similar to the Japanese organization described by W. G. Ouchi, *Theory Z* (Reading, MA: Addison-Wesley, 1981).

12. Levering, Moskowitz, and Katz (1984), pp. 53–54.

13. J. O'Toole, *Vanguard Management: Redesigning the Corporate Future* (Garden City, NY: Doubleday, 1985), pp. 95–97.

14. A. M. Webber, "Consensus, Continuity, and Common Sense: An Interview with Compaq's Rod Canion," *Harvard Business Review*, 4 (July–August 1990), pp. 116–17. It appears that Eckhard Pfeiffer, Compaq's new CEO, is carrying on the Rod Canion tradition of running the company as a collaboration culture. When questioned Pfeiffer said that the main change at Compaq concerned competition in the fast-moving computer market, not in the culture. Levering and Moskowitz, (1993), p. 65.

15. Webber, p. 118.

16. C. Hymowitz, "Which Corporate Culture Fits You?" *The Wall Street Journal* (July 17, 1989).

17. Levering, Moskowitz, and Katz (1984), pp. 65–66.

18. Harmon and Jacobs, p. 139.

19. Levering, Moskowitz, and Katz (1984), p. 124.

20. J. Case, "Collective Effort," *Inc.* (January 1992), pp. 35, 42–43.

21. Harmon and Jacobs, p. 145.

Chapter Five

The Competence Culture

The competence culture emerges from the socialization base of educational organizations, particularly the university. Educational institutions play a significant formative role in our lives. Indeed, a college education is nowadays a prerequisite to most gainful employment in the United States and many other countries. The prototype of the university as an organization serves as the third fundamental framework that leaders rely on to create the third core culture, the competence culture.

The university is a world of expertise and advancement of knowledge; the emphasis on these two pervades the competence culture. A competence culture is one of ideas, concepts, and technologies, in which scientific thinking and theoretical pursuits hold sway. Like the university, the competence culture thrives on encouraging intellectual and technical capability.

The competence culture is based on the achievement motive, discovered by D. C. McClelland in his research on individuals and societies, and defined as man's need "to compete against a standard of excellence." According to McClelland, people with a strong achievement motive act in four characteristic modes.

- They set realistic, not impossible, goals.
- They prefer work situations in which they take personal responsibility for achieving those goals.
- They want feedback on their own performance.
- They show initiative in researching their environment, traveling, trying new things, and searching for new opportunities.[1]

The need to achieve has to do with accomplishing more and doing better than others. People who value achievement try to set new standards and to improve on or surpass prior levels of accomplishment. They keep pushing to increase their level of competence.

Competent means to be "well qualified; capable; fit." Interestingly, the root of competence is from the Latin word *competere* which means to "compete." A competence culture is permeated with the value of competition against a standard of excellence. And from a social institution standpoint, the university (and education generally) is where one develops the competencies that one needs in order to achieve and compete. The university is a natural prototype on which to rely if one wants to create and build a competence culture.

Accordingly, the competence culture's way to success is to create an organization that has the highest possible level of competence and to capitalize on that competence.

Success in a competence culture means to be superior, or the best—to win by having a product, service, process, or technology that is unequaled in its marketplace or sphere of endeavor. The organization's customers or constituents can't go wrong with an organization that offers the best product or service. There is no need to go anywhere else.

Bell Laboratories, the research and development arm of AT&T, is a competence culture. It is one of the most prestigious scientific research organizations in the world, chiefly in the field of telecommunications. Among Bell Labs' inventions are the transistor, sound movies, modular phones, and the display on digital watches. It first synthesized Vitamin B1. In 1978 two Bell Labs scientists, Arno Penzias and Robert Wilson, were awarded Nobel Prizes for the discovery of faint background radiation, which confirmed the theory that the universe was created billions of years ago by a "big bang." In August 1983, Bell Labs received its 20,000th patent. A standard of excellence definitely makes things tick at Bell Laboratories. It is an exciting place to be because the people work on the leading edge of their disciplines. One can witness or be a part of technological breakthroughs.

> Bell Labs provides the closest approximation to an academic environment that you can find within the walls of a commercial establishment. It's a collegial atmosphere. Scientists work in labs the way they would at a medical research institution or advanced engineer-

ing school. Various disciplines are represented at Bell Labs: electrical engineering, chemistry, biology, computer science, physics, psychology.[3]

In addition to research and development organizations, many consulting firms, accounting firms, think tanks, and engineering construction firms are competence cultures.

When Father Theodore M. Hesburgh became president of Notre Dame University at age 35, he was determined "to point it to the top" by concentrating on excellence:

> I wanted Notre Dame to be a great university and also to be a great Catholic university. The first is easier to do than the second, for there are many great universities; but there has not been, since the Middle Ages, a great Catholic university. So we were working in an unknown field, creating something new. In the university world, we need to compete with the best, and it is pretty obvious how you go about trying to do this. A great university is fairly simple at its core: it must have a great faculty, a great student body, and a great facility.[4]

The competence culture is thus permeated with assumptions about achievement and being the best at what it does. It naturally pursues excellence, a management concept that has received great emphasis in the last decade.[5] Pursuing excellence is essential to the competence culture, but it is not necessarily the best way for all organizations to accomplish success.

Much of the focus of those who write about organizations has a technical and scientific emphasis to it.[6] Of all the four core cultures, the competence culture has generated the greatest number of developmental works on the study and improvement of organizations. This is to be expected considering that most writers themselves belong to competence cultures such as universities and consulting firms. To some extent, it is like watching a baseball game; what you see depends on where you are sitting.

This culture gains its uniqueness by combining possibility with rationalism. What might be and the logic for getting there are what count. A great deal of ingenuity and creativity is present in organizations belonging to this culture, but life is more impersonal than personal. Individuals work primarily to serve the concepts and theories of the organization. Whereas the control culture strongly values organizational systematism, the competence culture values conceptual systematism. More inventions, patents, creative theories, scientific discoveries, and new technologies and services come from this culture than from any other.

Lincoln Electric in Cleveland, Ohio, owns a huge factory (1.7 million square feet) that has a tunnel connecting the company's administrative and marketing offices with the plant itself. Etched in large stainless-steel letters across the entrance to this tunnel is the company's motto: "The Actual Is Limited, The Possible Is Immense." Lincoln Electric is the world's largest manufacturer of arc welding products. The company is known for its constant innovations and its highly competitive internal culture where people are given the incentive to achieve more and more. The average incentive bonus paid out to production workers in 1990 was $21,000.[7]

Pioneering organizations are often competence cultures. Cray Research is the world's leading designer, producer, and servicer of supercomputers for customers in the scientific community. Its computers, the largest of which is capable of making one billion calculations a second, have become indispensable to those who do state-of-the-art work in nuclear physics, oil exploration, weather prediction, aircraft design, and weapons research. Its fundamental strategy is

> innovation—continuously creating new generations of supercomputers that at times threaten to put Cray out of its "old" business by making obsolete the pioneer machines of the previous year. Cray leads in a *niche market* . . . [companies] whose requirements are so sophisticated and need for having the fastest supercomputer so great that only the best, fastest, and most powerful will do . . . Cray competes on the *value* of its product. It produces and markets among the most expensive computers in the world—selling them to price-conscious government agencies, research institutions, and universities—because they get the work done cheaper and faster, and make possible research that could not be done before. Cray has an organizational culture that transcends wealth creation and business administration. Founded by five men who sought to escape the administrative bureaucracy of a larger company and specialize as technical purists, the company's shared values and commitment to a small company spirit require few rules as Cray celebrates autonomy, common sense, and fun . . . More importantly, Cray shares the common interests and needs of leading-edge scientists—the best for the brightest.[8]

Technological innovation and superiority is Cray's way to success. From the outset, Cray Research invested nearly $21,000 per employee in research and development, a figure nearly four times the average of other computer companies such as IBM, Honeywell, Control Data, Apple, and Tandem. *Business Week* ranked Cray Research first in investment per employee in R&D among major US companies that spend at least 1 percent of their sales on R&D. Cray Research believed that maintaining "continued technological superiority was central to its mission and that its leading-edge customers—for the most part sophisticated scientists in their own right—would pay the premium for continuing innovation."[9]

Another notable example is Merck, the world's largest manufacturer of prescription drugs. Merck spends $1 billion a year on research. According to Dennis Schmatz, a research scientist at Merck, "Merck is the closest thing you're going to get to academic research in industry."[10]

A competence culture keeps building expertise because it accomplishes its desired objectives by knowing more about something than other organizations. Knowledge and information are fundamental. Professionals are attracted to this culture because it reinforces their feeling that they have a professional calling for what they do. Achieving technical ideals and gaining preeminence motivate them. Loyalties to one's profession are often stronger than to the organization.

Competence cultures are frequently meritocracies. Merit is the central value that you have to earn to be held in high esteem. To deserve esteem you have to demonstrate your competence. Talk is cheap; there is a lot of "show me." Formalities and emotional considerations take a back seat to demonstrated performance and proven accomplishment.

Predictably, the culture emphasizes doing things well. Success is intimately tied to being the best and achieving superiority, so people are encouraged, even compelled, to do the best they can. Nothing is taken for granted. Levels of accomplishment are simply plateaus or standards to surpass. This is a "stretch" culture in which products and services can always be improved upon. The individual and the organization succeed when they adopt and follow through on this belief. People in the competence culture are often more intense, high strung, and present a sense of urgency than people in other cultures. They have a hard time celebrating or feeling satisfied about what they have done and don't rest long on their laurels. A high level of discipline pervades. There is an almost constant emphasis on efficiency, of getting more out of less.

The founder of Automatic Data Processing (ADP) in Roseland, N.J., Henry Taub, works day and night. ADP's customers require accuracy and on-time delivery which causes unusual intensity of purpose, of commitment, of day-to-day execution. While this intensity generated record sales and earnings for more than 30 years, people who work at ADP do not find it a restful place.

The CEO sets the pace for the entire organization. ADP

> "does not believe in trade-offs. Its people are expected to achieve short-term performance, long-term performance, and the development of a strong organization. This attitude generates adrenaline and creates tension, but it also results in a higher level of corporate achievement. Perhaps the lesson is that the dissatisfied organization, the one with high standards that can never be fully met, is the one that will create constructive tensions that lead to long-term success."[11]

A competence culture values competition for its own sake even though it is not necessarily more competitive than other core cultures. There is a love of challenge; people like to be told that "it can't be done." New York's Citicorp, according to one young Citibanker, "is the only place in town where the elevators are crowded at 11:30 at night. Maybe we're just sixty thousand workaholics." Citicorp attracts a distinct breed: people who relish action, thrive on competition and love pressure. Stuffy types need not apply. For Citicorp is "hell-bent on changing the nature of banking, and it wants people who prefer living in the eye of a hurricane." Citicorp is a trendsetter in banking with its invention of certificates of deposit, launching of automated teller machines, establishment of an electronic banking satellite network, and leadership in interstate banking.[12]

Microsoft, the software developer, is a competence culture. William H. Gates III, its young founder and CEO, is an intensely competitive technical wizard who handles people abrasively. Jay Blumenthal, director of program management says: "One of the things that characterizes Microsoft is insecurity. People never believe that they've got the best product. It's always the fear that somebody is going to come up with a better product. Bill [Gates] is at the top and he always wants it to be better and everybody has picked that up through the years.[13]

Craftspeople are known for being the best at what they do. In a competence culture, people take pride in looking at problems in depth, doing the job in an exhaustive way, and approaching work in a thorough manner. No hip-shooting goes on; the toughest problems are dealt with carefully. It is a big mistake to make a claim not backed by logic or facts. Just as it is attracted to competition, the competence culture is attracted to problem solving. Tackling complex problems is inherently enjoyable. They give people a chance to test their mettle and capability. Intel Corporation people describe the company's style as one of "snake biting," meaning that they are constantly looking for problems ("snakes") to solve. David House, who runs the company's development systems operation, says: "When a good, solid problem is discovered, we study the daylights out of it. How big is it? What color is it? How much does it weigh? Everybody studies, looks, jabbers about the problem and admires it, like putting it on a pedestal. Finally somebody breaks the code, grabs it and stomps it to death. Everybody cheers."[14]

Four Seasons Hotels, Ltd., is an excellent example of craftsmanship. Issy Sharpe, chairman of the board, personally oversees much of the initial interior design of new hotels on the drawing board. The hotels themselves are works of art. The company goes to great lengths to bring in the best hoteliers in the business. They attend to every detail of providing accommodations to customers. The chefs are among the world's finest. Executives dedicate great attention and energy to developing services that are unique in the hotel industry. Four Seasons has developed nearly 700 standards for people to follow; these standards are constantly upgraded. The company's strategy for success is simply to be the very best hotel company in the business.

Regardless of the kind of enterprise, competence culture organizations appeal to the needs of their customers and constituents for unequaled service, products, and information. However, customers and clients should expect to pay more.

Expediency, complacency, or taking the easy path do not work in this culture. Nor is it a good idea to be illogical, irrational, or slapdash. Unlike people in a control culture, people in a competence culture do not like work that is a piece of cake because it is no fun and lacks stimulation.

Competence culture people are analysts who value formal logic and deduction. They look at things from a skeptical viewpoint and emphasize finding the single best way. Empiricism and experimentation are highly valued. People expend considerable effort coming up with models and formulas, methods and plans; the more scientific the solution, the better. Creativity is highly valued only when channeled in directions that leaders believe will benefit the organization. People tend to be more impersonal with one another. Concepts, theories, ideas, and possibilities have the highest priority.

Competence culture people crave feedback from one another and from their environment. They want to know whether they are the best, so the level of market research is high. People are open to corrective action, and considerable monitoring and tracking occurs. Competence culture people value flexibility and adaptability.

LEADERSHIP, AUTHORITY, AND DECISION MAKING

Leaders in the competence culture are standard setters. They build a vision for their organization and enlist their peoples' commitment to that vision primarily through assertive, convincing persuasion. In turn, people follow their leaders because they believe in being the best. They work to establish their own and their organization's preeminence. Leaders in this culture are often visionaries and architects of systems. They call upon their people to give their all because the organization has such challenging and unique objectives to attain. Leaders focus on new products, services, markets, businesses, and opportunities.

Robert L. Swiggett, CEO of Kollmorgen Corporation in Stamford, CT, exemplifies a competence culture leader who communicates his vision for the company.

> I want you to visualize a very bright white light . . . Try to visualize the largest diamond you've ever seen coming in from the right. The

diamond moves over where the light is, and the light is absorbed into the heart of that brilliantly cut three-inch diamond. Visualize the diamond slowly rotating, and, every time one of the facets of the diamond lines up with your eyes, it scintillates . . . [a] blue-white light coming out of the heart of that diamond. Think about the white light in a verbal sense. Think that it means number one. First. Absolutely first. Think about the diamond with the white light in the center that means first. I want to use that physical image to represent our vision of what we're trying to do in our company . . . How about first to market with the best technology? . . . How about first to the market with the best proprietary technology in growth markets? . . . It all starts with the idea of being first. First with the best. And what we try to do in our business, in some way, is to measure everything we do against this big diamond in the sky, being the first with the best.[15]

The belief is that ideas and concepts ultimately move people. They bring about the raising of armies and the outcomes of elections. They mobilize people to do their very best and to achieve more. Winston Churchill and Abraham Lincoln would have been natural competence culture leaders. Leaders are often very good at building conceptual frameworks; at developing prototypes, pilots, and models; and at designing plans for change or development.

Leaders emphasize incentives and differential reward. They foster individual and group competition. The better you can be in this culture, the more you will be rewarded. However, people are perceived to have essential differences in talents, abilities, and willingness to work. These differences are acknowledged and rewarded accordingly.

Competence culture leaders are strategists by nature. They take the long-range view, looking down the road and striving to anticipate every contingency. They accord considerable importance to strategic planning. Once plans are in place, leaders ensure that subsequent activities conform to them. Strategy and its implementation ensure the organization's future, but possibilities are never overlooked. Everyone in this culture looks to the future. The real excitement lies in what's ahead and what can be created that is better.

Seymour Cray "was obsessed by a vision of technological achievement, scientific purity, and quality," and set out in 1972 to design the best supercomputer. A compatriot, Les Davis, said later: "We really didn't know exactly what, if anything, we would produce. We might have become an engineering think tank or we might have assembled the fastest computer—but we were in the business of advancing the state of the art!"[16] Cray's team worked four years and spent millions of dollars before it sold a single computer. But Seymour Cray and his associates remained obsessed with their goal of creating more powerful supercomputers and let nothing stop their progress. A person in the industry jokingly remarked, "The first supercomputer was like winning a Nobel Prize. But these guys are not satisfied unless they win the prize every year."[17]

Competence culture leaders are frequently never satisfied. There are always more things to do, bigger hurdles to overcome, higher mountains to climb. Leaders are often very hard taskmasters and exacting in their expectations of others.

Ray Jones (a pseudonym), CEO of a waste management machinery manufacturing firm, deeply values competence and continued achievement. For him, the past is gone, the present is everchanging, and the future keeps beckoning. He has created and patented a can densifier that crushes and packages aluminum cans and is in the process of personally patenting another new product. Jones thrives on challenges. He continually focuses on how the organization can do things better, more efficiently, and more productively. Now in his mid-60s, he still arrives at 6:00 A.M. every day and doesn't leave work until 6:00 or 7:00 P.M. He runs a tight ship, and his subordinates have great admiration for his tireless energy and dedication to being the best he can be.

In this culture, management is a rational activity: good thinking and good judgment take the organization ahead. Information is very important. More than the other three core cultures, competence cultures are voluminous information gatherers. Hard analysis must occur before sound decisions can be made. Good results don't just happen. They have to be thought through and carefully planned. Leaders must anticipate every contingency, so they rely on experimentation and empirical testing. If people aren't sure about something, they build prototypes and test them out before making a final decision.

At the same time, this culture is under considerable pressure to move ahead. Leaders often have a single purpose and they prefer to make decisions and to push on. The more efficient the decision-making process, the better. Decisiveness is valued. There is no place for wavering; you either know or you don't know. If you don't know, then you must have a plan of action to determine how you will know and how long it will take you. Goals and objectives must be crystal clear.

Expertise power counts in a competence culture. If you are *the* expert on something, you will have the most power concerning that something. So long as it is applicable, generalized expertise will give you widespread power.

STRUCTURE AND RELATIONSHIPS

Structure is less important in the competence culture. Leaders are not wrapped up in it and take the attitude, "let's organize and reorganize in whatever way it takes to get us there." Generally, concepts and conceptual systems hold sway more than organizational structure. The structure that fosters the implementation of concepts is what gets put in place. Demonstrated competence, efficiency, and productivity also drive structure and policy. The competence culture is one of natural improvisation. The actual structure varies according to the technical, scientific, and technological needs of the organization. The structure that best fits the nature of a competence culture is the matrix structure, one in which expertise is given structural reality.

Innumerable task forces are common. They fit well into the competence culture because they can be formed, deployed, and disbanded at will. A competence culture generates timebound and unique special issues or temporary projects. It does not emphasize assigning people to fixed tasks or fixed groups; people rotate in and out of both as the need arises.

Relationships in this culture are much more task oriented and impersonal than people oriented or emotional. Ideas and visions are what tie people together. Accordingly, people connect with one another on one task and with other people on another task.

The human and emotional elements are subordinated to the conceptual elements of the organization. People negotiate a lot with one another. You make headway with people by impressing them with what you know or have already achieved; the more you know or have achieved, the higher the probability that you will get what you are after in your negotiations.

Internal relationships are often competitive. Microsoft Corporation's Windows NT software development project is a clear, even harsh, example of how competitive life can get. In 1988 CEO Bill Gates brought in David Cutler, formerly of Digital Equipment, to head up Microsoft's NT software development project, a project of great importance for Microsoft. It was believed that Cutler's style meshed well with the confrontational environment that thrived at Microsoft. Cutler pitted software engineers

> against one another in a deliberately cacophonous search for answers to programming questions that [grew] more complex with each additional line of code (eventually the NT program reached the equivalent of 100,000 single-spaced pages of type) . . . [coders and testers] waged some of the biggest battles over reliability. Every night the testers would subject the fresh NT code to ordeals on 200 computers, stresses that would rarely occur in the real world. Testers might ask NT to read and write, continuously and simultaneously, in separate files all night long. At the same time they might instruct the program to create a graphic on screen and then wipe it out, over and over again. A report would be generated in the morning, and any computers that couldn't complete the test—that crashed or halted—would be investigated. On a great night, only 2 percent of the computers would fail. . . .

As the project wound down after nearly two years, Charles Whitmer, a graphics programmer, said, "People are really worn out . . . angry, tired and burned out."[18]

The competence culture breeds technical specialists. In contrast to the control culture, organizational functions are built around technical specialties, special disciplines, or unique technologies. More people from a wider range of technical disciplines are brought into work in a competence culture than in the other three core cultures.

Roles and jobs are much more broadly defined and shift more frequently than in control or collaboration cultures. In order to achieve the goal, there is no hesitation to redefine a job or role in the organization. Titles change a lot. The division of labor and the deployment of talent and specialist expertise are discussed frequently.

Relationships are flexible and fluid in the competence culture. Organizations adapt and change, and readily take corrective action. People must be open and ready to "go with the flow." Projects and subprojects abound; depending on circumstances, people move frequently from one project to another.

In general, relationships in a competence culture are characterized by individual freedom. People feel a strong need for autonomy and the opportunity for individual achievement.

STAFFING AND PERFORMANCE MANAGEMENT

Competence cultures recruit people who can make the greatest contribution to the vision of the organization. Hiring is based on prior demonstrated competence and achievement, and graduate degrees often carry great weight. The culture treasures education; some competence culture organizations will not consider people for membership who do not have a graduate degree from a prestigious university. Technical and scientific disciplines proliferate. According to C. Kumar N. Patel, a physicist at Bell Laboratories: "Every person hired by Bell Labs comes with technical excellence. But doing science or technology is more than working in your own lab or subgroup. It requires convincing your colleagues of the importance of your work."[19]

The success of a person hired into a competence culture is contingent upon demonstrated performance. While this principle is true for every core culture, it is particularly true for the competence culture. Technical mastery is very important. The more you know about your field of expertise, the better you do. The rule is often "up or out"; while this is the extreme, it represents the attitude. The culture has a great deal of mobility. Success is tied to what a person merits. Robert N. Noyce, cofounder of Intel Corporation, says: "There is no resting on your laurels, because you will get wiped out next year if you sit back."[20]

This is a culture for superstars. If you are the best, you will find this culture receptive.

> Behind locked doors on the fifteenth floor of Citicorp's Park Avenue headquarters sits a room devoted to what's called "Corporate Property." Pinned on a board . . . are the photos and biographies of about 75 managers, considered to be the bank's up-and-coming superstars. The only people permitted entry . . . are the bank's top two dozen senior executives. They make sure the superstars get special attention and job transfers to acquire a broad-based experience at the bank.[21]

The competence culture is often a management-by-objectives (MBO) culture. It puts a lot of stock in achieving agreed-upon goals, and it closely measures and analyzes performance. Achievement-oriented people need ongoing knowledge of results. They find that this high feedback culture emphasizes ongoing monitoring of what is working or not working. Intel's Noyce says, "High achievers love to be measured, when you come down to it, because otherwise they can't prove to themselves that they are achieving."[22]

Ongoing training and education is prevalent. Bell Laboratories has an in-house education center that offers courses ranging from accounting to computer science. A two-year course provides employees the equivalent of a master's degree in some courses. The courses are held either on the premises or at outside schools. The company pays for employees' tuition.[23]

Conflict in competence cultures is managed by logic, reason, and debate. The argument that has the greatest command of facts or the most competent line of reasoning comes out on top.

Rewards are predicated primarily on performance and expertise. Compensation differentials are considerable. An individual who is a recognized expert or proven performer can make a lot of money. Those who fail in the stiff competition make less and sometimes are washed out of the system. Paul Cook, CEO of Raychem Corporation, another competence culture, describes Raychem's bonus system:

> Some companies spread bonuses quite evenly among group members. We have a different approach. Typically within a division there are significant differentials based on performance. Having a big

spread causes some unhappiness. But it also creates drive, because I think people respect how we evaluate their contribution.[24]

The happiest people in a competence culture tend to be analytical—the broad conceptualizers and technical specialists, independent thinkers and complex problem solvers. They enjoy finding new challenges and solving them. They are creative and possibility oriented, like using the scientific method to get things done, and are logical and crisp thinkers. They enjoy intellectual repartee and their feelings are not hurt easily. As professionals they value personal discipline. By nature, they are competitive.

WHERE DOES A COMPETENCE CULTURE FIT?

The competence culture is naturally suited to enterprises that create market niches or come close to doing so. Its primary nature is one of creating distinctive products or services and bringing them to customers or clients who want and can pay for them. This culture seems to be in the business of creating its own marketplace. Its niches are usually circumscribed, but they can be quite large, at least in terms of revenue generation—specialized computer hardware and software applications come to mind.

Enterprises in the competence cultural framework have services or products that offer distinction or uniqueness and, curiously, help people to have more freedom. Examples include state-of-the-art or one-of-a-kind services or products.

Appendix A to Chapter Five
STRENGTHS OF THE COMPETENCE CULTURE

It has high performance standards.
It has a high continuity of service.
Institutional wisdom and its preservation are important.
It offers considerable technical expertise.

Great achievements come from this culture—new inventions, technologies, services, and products. When knowledge is advanced, it happens in this culture.

It is goal oriented and results oriented.

You can't go wrong with its products, services, or technologies.

It is future oriented and possibility oriented. It often sets trends.

It is a creative and exciting place to work.

It is visionary. It values going beyond what has already been achieved.

It puts much more into research and development than other cultures.

It places a high value on professionalism.

It emphasizes merit and demonstrated performance.

Life gets to a "high pitch."

Discipline is present and emphasized.

It is an efficient and productivity oriented culture.

It values craftsmanship.

The reward system is incentive oriented. An individual can make a lot of money.

It accords considerable importance to strategy and planning.

Decision making is thorough, considered, and systematic.

An individual can stand out.

It is good at adapting and changing.

Ongoing training and education are prevalent.

Appendix B to Chapter Five
WEAKNESSES OF THE COMPETENCE CULTURE

In excess, this culture leads to technical or expert tangents. The organization slips into directions that lack viability or pragmatism. It loses sight of the need to stay focused on the application of ideas.

It loses sight of the human element and takes people for granted. Good (not superior) people are passed over. People are treated insensitively; personal concerns become annoyances. Bright, capable, but less educated people are screened out, resulting in the loss of valuable resources.

Values and subjective views are prone toward getting screened out or ignored.

People may overplan and overanalyze.

It is too emotionally controlled.

In excess, it gets so involved in the world of ideas that it fails to appreciate the real world of people, time, individual weaknesses, or prejudices. It risks thinking that the model represents the real world, instead of an attempt to simulate reality.

People feel that leadership is never satisfied and that they are always underperforming.

It is too tough on people, too much the taskmaster. In the extreme, people feel like they cannot make a mistake or say that they don't know or can't do something. There is a great deal of pressure to keep pushing on and to refrain from admitting that you have difficulty keeping up.

Winning takes on too much prominence.

In the extreme, "win-lose" behavior goes on. What could be a "win-win" situation is overlooked or not considered.

It advises against collaboration and teaming because that interferes with individual recognition and prominence.

If overdone, people are overly pressured, overworked, and stressed out. People feel like they can't relax, and they don't celebrate enough.

Capable people who are less effective communicators feel unappreciated.

In excess, it makes people feel constantly insecure.

Generalists are not developed or encouraged.

NOTES

1. Quoted in Thomas R. Horton, *What Works for Me: 16 CEOs Talk about Their Careers and Commitments* (New York: Random House, 1986), pp. 403–4.

2. R. Levering, M. Moskowitz, and M. Katz, *The 100 Best Companies to Work for in America* (Reading, MA: Addison-Wesley, 1984), p. 27.

3. Ibid., p. 28.

4. Horton, p. 154.

5. T. J. Peters and R. H. Waterman, Jr., *In Search of Excellence: Lessons from America's Best-Run Companies* (New York: Harper & Row, 1982); T. J. Peters and N. Austin, *A Passion for Excellence: The Leadership Difference* (New York: Random House, 1985); T. J. Peters, *Thriving on Chaos: A Handbook for a Management Revolution* (New York: Alfred A. Knopf, 1987); J. O'Toole, *Vanguard Management: Redesigning the Corporate Future* (Garden City, NY: Doubleday, 1985); D. K. Clifford and

R. E. Cavanagh, *The Winning Performance: How America's High-Growth Midsize Companies Succeed* (New York: Bantam Books, 1985); F. G. Harmon and G. Jacobs, *The Vital Difference: Unleashing the Powers of Sustained Corporate Success* (New York: AMACOM, 1985).

6. See the following works by Peter Drucker: *Innovation and Entrepreneurship: Practice and Principles* (New York: Harper & Row, 1985); *Management: Tasks, Responsibilities, Practices* (New York: Harper & Row, 1973); *Managing for Results* (New York: Harper & Row, 1964); *Managing in Turbulent Times* (New York: Harper & Row, 1980); *The Changing World of the Executive* (New York: Times Books, 1982). See also M. Walton, *The Deming Management Method* (New York: Dodd, Mead, 1986).

7. R. Levering and M. Moskowitz, *The 100 Best Companies to Work for in America* (New York: Currency/Doubleday, 1993), p. 236.

8. Clifford and Cavanagh, pp. 171–72. Between 1987 and 1990, Cray lost three key people, including its founder, Seymour Cray, who went off to start another supercomputing firm. Hit by financial hard times, the company nonetheless is pushing ahead with considerable intensity and is dedicated to restoring its former financial successes.

9. Ibid., p. 174.

10. Levering and Moskowitz (1993), p. 280.

11. Clifford and Cavanagh, pp. 190, 195, 196.

12. Levering, Moskowitz, and Katz (1984), p. 45.

13. Levering and Moskowitz (1993), p. 287.

14. Levering, Moskowitz, and Katz (1984), p. 154.

15. Quoted in J. M. Kouzes and B. Z. Posner, *The Leadership Challenge: How to Get Extraordinary Things Done in Organizations* (San Francisco: Jossey-Bass, 1987), pp. 106–7.

16. Clifford and Cavanagh, p. 185.

17. Ibid., pp. 185–86.

18. G. P. Zachary, "Climbing the Peak: Agony and Ecstasy of 200 Code Writers Beget Windows NT," *The Wall Street Journal* (May 26, 1993).

19. Levering, Moskowitz, and Katz (1984), p. 29.

20. Ibid., p. 155.

21. Ibid., p. 46.

22. Ibid., p. 153.

23. Ibid., p. 28.

24. W. Taylor, "The Business of Innovation: An Interview with Paul Cook," *Harvard Business Review*, 2 (March–April 1990), pp. 100–101.

Chapter Six

The Cultivation Culture

An organization in Boulder, Colorado, has put together a credo to which it would like its people to subscribe.. The following is an excerpt from its statement of beliefs:

We believe in marketing and selling healthful and naturally oriented products that nurture people's bodies and uplift their souls. Our products must be superior in quality, a good value, beautifully artistic, and philosophically inspiring.

We believe in the dignity of the individual, and we are totally committed to the fair, honest, kind, and professional treatment of all individuals and organizations with whom we work.

We are committed to the development of our good people by identifying, cultivating, training, rewarding, retaining, and promoting those individuals who are committed to moving our organization forward.

We believe in fostering an environment which promotes creativity and encourages possibility thinking throughout the organization. We plan our work to be satisfying, productive, and challenging. As such, we support an atmosphere which encourages intelligent risk-taking without the fear of failure.

Our role . . . is to play an active part in making this world a better place by unselfishly serving the public. We believe we can have a significant impact on making people's lives happier and healthier through their use of our products.[1]

What kind of organization is this? Which would you choose— (1) a business organization? (2) a religious organization? or (3) a governmental organization?

If you selected answer 1, you were correct. But it is understandable if you picked answer 2. The credo belongs to Celestial Seasonings, Inc., a multimillion dollar enterprise that makes and sells herbal tea. It is an example of a cultivation culture organization.

The fourth core culture has its socialization base in *religious organizations*. The church or synagogue or mosque are powerful formative influences; some would say that religion is *the* most powerful of influences for most of us. Religious organizations are dedicated to furthering the human spirit, inculcating ethics and values, and uplifting humankind to a higher plane. How systems of belief operate provides a fundamental framework for culture formation and implementation.

The cultivation culture is one of *faith*. It heralds a system of beliefs or expectations that the organization and its people will accomplish what it deems *valuable*. What is important is that people believe they are making progress toward a higher order or worthwhile level of accomplishment. This is the culture of meaning and transcendence. *Transcend* means to go beyond the limits of, to exceed, and to be extraordinary. In a cultivation culture, people work hard and dedicate themselves to accomplishment because they believe in a higher order of values and continued self-development. This culture trusts unquestioningly in success, in its people, and in the organization. When the culture succeeds, it does so because its people expected success and capably accomplished the things they considered valuable.

Celestial Seasonings is not only an herbal tea business but also an organization dedicated to making the world a better place, nurturing people's bodies, and uplifting their souls. It is the purpose that counts; the company's products are simply an expression of this purpose.

The cultivation culture is a *purposive* culture. It acts the way it does because it has a purpose. It has resolution and determination; it succeeds because its people will it so. People believe deeply that what they do is valuable to themselves and others. When the mission of the organization is achieved, it has fulfilled a purpose not only for the members of the organization but also for its customers or constituents.

Max DePree, Chairman and CEO of Herman Miller, Inc., the furniture maker in Zeeland, Michigan, offers this philosophy of management:

> In addition to all of their ratios and goals and parameters and bottom lines, it is fundamental that corporations have a concept of persons. In our company this begins with an understanding of the diversity of people's gifts. Understanding the diversity of our gifts enables us to begin taking the crucial step of trusting each other.

DePree works very hard at inculcating trust into the Herman Miller way of operating. He prefers "roving leadership" to hierarchical leadership and "covenantal relationships" to legal or contractual ones. To DePree a covenantal relationship is

> based on shared commitment to ideas, to issues, to values, to goals, and to management processes. Words such as love, warmth, and personal chemistry are certainly pertinent. They fill deep needs and they enable work to have meaning and to be fulfilling. Covenantal relationships enable corporations to be hospitable to the unusual person and to unusual ideas.[2]

The energy and the vitality of the cultivation culture come from commitment and the individual and collective fulfillment of valuable and worthwhile purposes.

While D. C. McClelland researched and articulated the motives for the first three core cultures, the work of Abraham Maslow provides the cultivation culture with its individual motivation base.[3] Maslow dedicated most of his career to refining his discovery of people's need for "self-actualization" or growth. Maslow argues that people are motivated to grow in order to enrich their lives and expand their horizons. Satisfaction is accomplished through a complex process that leads to the realization of capabilities or ideals. The fundamental human motive has to do with creative living, unselfish love, unbiased understanding, personal fulfillment, and personal completeness. Maslow's self-actualized person has the following characteristics:

• Acceptance of self, others, and the natural world.
• Spontaneity.

- Task orientation rather than preoccupation with self.
- Independence.
- Vivid appreciativeness.
- Spirituality that is not necessarily religious in a formal sense.
- Sense of identity with mankind.
- Democratic values.
- Recognition of the difference between means and ends.
- Creativity.[4]

Notice the close connection between the growth motive articulated by Maslow and the belief and value orientation of religious organizations.

Shorebank Corporation in Chicago exemplifies how a cultivation culture organization dedicates itself to expanding the horizons of others. Shorebank Corporation is a multifaceted for-profit business created specifically to bolster the fortunes of South Shore, a community that borders downtown Chicago. In 1972, the South Shore was in a terribly deteriorated condition. That same year the South Shore National Bank, the area's primary financial institution, requested permission to relocate downtown. The request was denied and the bank was sold to a group of investors headed up by Ronald Grzywinski. At the time of the sale, the bank's book value was $3.2 million.

By 1992, the bank's deposits alone totaled $200 million and the bank had recorded its 16th consecutive year of profitability while lending money to South Shore residents who otherwise might not have obtained loans. Ron Grzywinski and others accomplished this impressive growth and profitability by dedicating themselves and Shorebank Corporation to expanding the horizons of the South Shore community and its residents. Grzywinski said, "We had an idea. We had a vision. We knew that there had to be a better way to rebuild the cities of America for the benefit of the people who lived there."[5] Grzywinski and Shorebank's management team accomplished their goals by forming a bank holding company designed to revive the South Shore neighborhood by reinstalling credit, rehabilitating self-confidence, and reestablishing a functioning market economy. Under the holding company umbrella are a bank, a real estate development subsidiary, an investment company for minority small business enterprises, a nonprofit affiliate that creates housing and jobs, and a consulting advisory affiliate for companies involved in economic development. Thanks to the innovative efforts of Shorebank Corporation, South Shore's property values have risen rapidly, thousands of rental units have been renovated and leased, new shopping centers and apartment buildings have sprung up for the first time in decades, dozens of new businesses have received organizational support, and thousands of local residents have benefited from educational and job placement programs.[6]

Analog Devices, the digital components and systems company in Massachusetts is a high-tech company whose engineer founder, Ray Stata, doesn't mind being called a "humanist." Individual employees at Analog Devices have a great deal of freedom. The company manual states that "While acknowledging the necessity for policies, procedures, and basic management control systems, we attempt to hold these to a minimum and invest more effort in developing sound human judgment. This reflects our belief that individual judgment is generally more reliable than rules and regulations." Stata challenges people to perform at the limit of their potential, and he doesn't believe something can be done with too many rules. "We try to break the procedural syndrome whereby managers impose themselves on others," he says. "We're not trying to eliminate all hierarchy but to cut the counterproductive values generated by the rules. The greatest limitation in traditional organizations is that people farther down the ladder always consider themselves less valuable and less creative human beings."[7]

The cultivation culture's way to success is to create and provide the conditions whereby people within the organization grow, develop, and strive to accomplish the highest-order purposes possible. It does this by cultivating its people. This culture is in many ways a farming culture, organic and highly adaptive. It tills the soil of its internal environment, plants the seeds for growth, and trusts in nature to take its course. As growth occurs and when the time is ripe, this culture harvests the ideas, energy, and output of its people.

Advanced Micro Devices in Sunnyvale, California, has a poster on the walls of its headquarters which shows a color photo of a bunch of asparagus, commemorating "The Age of Asparagus." The asparagus in the poster symbolizes AMD, which like the asparagus crop, was slow to nurture but then grew very rapidly.[8] Control Data Corporation in Minneapolis has a four-acre company garden just beyond the employee parking lots. Some 600 Control Data employees and their families grow tomatoes, corn, squash, and other vegetables. Each year the garden club presents a Golden Hoe Award to the employee who grew the biggest vegetables. The roof of a research facility supports a hydroponic greenhouse that produces lettuce sold at bargain prices to employees.[9]

Farms and churches put their faith in a higher order; they cultivate their soil/people; harvest/gather in their crops and ideas/activities/new initiatives. Both are growth enterprises that nurture and develop through care. They encourage and stimulate. Both strive to remove barriers (weeds/lack of understanding or opportunity) to growth.

The natural and automatic meaning of success for a cultivation is the fuller realization of potential. The pursuit of possibility makes this culture go. But unlike the competence culture, the cultivation culture is dedicated principally to people possibilities—unleashing people's potential and fostering individual and collective completeness. Work in this culture is to be an ennobling experience. Humanism is highly valued; what is important to people and society is given prominence as are feelings and emotions. Cultivation cultures strive to realize human ideals. How people feel about things is important. The organization exists more to serve the individual and society than vice versa. By contrast, control and competence cultures place the individual more in the service of the organization or its concepts. In the cultivation culture organization, the individual is the basic element. One employee at Esprit de Corp., the successful San Francisco-based sportswear company, said "it's a sin here not to realize your potential."[10]

In the early 1900s Maria Montessori of Italy developed the method of educating young children that bears her name. A Montessori school is a cultivation culture. The Montessori method teaches that children should be free to find out things for themselves and to develop through individual activity. Children learn how to learn by themselves. The focus of teachers is to create the conditions for child self-development and to cultivate the inherent potential of children in their care. At present there are more than 450 Montessori schools in the United States alone.

The cultivation culture is an uplift culture, striving to elevate people and society to a higher plane. It touches a vital core in the human psyche—the need to believe that a greater dawn can come, and that we can reach higher and expand our own and others' horizons. The promise of the future keeps people plowing ahead today. The future itself is a great adventure that may require adaptations; therefore, aspirations fill the agenda in this culture. Ideas, products, services, and processes that fulfill human aspirations and serve society's needs are cultivated and harvested.

Assumptions about change permeate life in the cultivation culture. Of all the four core cultures, this culture has the easiest time engaging in internal change and adapting to external circumstances. In this culture it is almost automatic that change, learning, and development go hand in hand. Carroll Skiba, an engineering consultant at Control Data, says that Control Data's ever-changing atmosphere is "like Minnesota weather. If you don't like it today, wait until tomorrow."[11] Control Data, like many cultivation cultures, is constantly involved in new fields and enterprises.

Many cultivation culture organizations advance a cause because they believe in making a contribution to humanity. John Gardner's Common Cause and the Sierra Club are examples. In the United States alone, at least 50 organizations are devoted to furthering international peace. Social change and mental health organizations are often cultivation cultures. Control Data was an early leader in corporate social responsibility. It built a production plant in a depressed area of north Minneapolis, hired people from the area, and made it into one of Control Data's most productive facilities. The company has engaged in other socially responsible activities. Its chairman, William Norris, defined Control Data's mission as "addressing society's major unmet needs as profitable business opportunities." The company's business strategy is based on that definition. "In a low-key way, Control Data sees itself as trying to save the world through computer technology."[12]

Organizations dedicated to aesthetics are often cultivation cultures: symphony orchestras, theaters, artistic organizations, and some entertainment, advertising, and media graphics enterprises. People here prize beauty, form, and harmony. They value and pursue grace, symmetry, and fitness. Again the key is to further realize potential, raise the human spirit, and broaden people's horizons. The three founders of Doyle Dane Bernbach International, Inc., (one of the largest advertising agencies in the United States), breathed a spirit into their agency based on respect for the individual and dedication to doing good work. The firm led a "creative revolution" in advertising through ads that had humor, warmth, and excitement. The ads didn't talk down to consumers and proved that sensibility could sell. Bill Bernbach once explained: "It is ironic that the very thing that is most suspect by business, that intangible thing called artistry, turns out to be the most practical tool available to it. For it is only artistry that can vie with all the shocking news events and violence in the world for the attention of the consumer."[13]

The cultivation culture is very often highly creative. Creating and inventing occur almost naturally. Inspiration is commonplace. The culture automatically encourages the free flow of ideas and initiatives. In many ways, creativity is what the cultivation culture is all about. The secret of 3M, an organization noted for its creative accomplishments, was described by *Forbes* as the company's ability to "nurture creativity."[14] 3M has nurtured and cultivated over 50,000 products that cut across virtually every industry. It expects one-quarter of annual sales to come from products that did not exist five years before.

> Two sayings typify 3M research. One is "Never kill an idea, just deflect it." The other, called the Eleventh Commandment: "Thou shalt not kill a new product idea." The burden is on those who want to stop research since the company has often found an application for many a seemingly off-the-wall idea.[15]

Self-expression is strongly encouraged in the cultivation culture—a natural "empowerment" culture. Recent management literature has given considerable emphasis to the empowering of people. Managers would do well to study this core culture for real-life examples of empowerment. In this culture, people are given considerable personal freedom and autonomy. It generously provides opportunities to explore, discover, and experiment, and it fosters reasonable risk taking. At Celestial Seasonings

> people can spread their wings. Helen Harring arrived there as a secretary in 1979, and she was later put in charge of automating the office with word processing stations. A janitor who thought money could be saved by having a fleet of trucks to transport teas to the marketplace instead of hiring outside truckers was given the job of organizing such a fleet.[16]

As is true in the other core cultures, the cultivation culture takes its cue for approaching customers and constituents from its internal way of operating. In essence, it strives to realize possibilities and potential more fully for its customers and constituents. It tries to fulfill a purpose for them, to expand their horizons, or to fulfill their aspirations. What sets this culture apart is its unique focus on people possibilities.

A cultivation culture is averse to anything that stifles growth, self-expression, or the realization of potential in people. Taking away the freedom of people to explore or to create is an example of this. Excessive control, proscription, or condemnation do not work. Telling people that they cannot make mistakes runs counter to what the cultivation culture is all about. Criticism is limited because it often has a diminishing effect on people. And because this culture values expansion, overemphasizing "what is" to the exclusion of "what might be" is also incongruent with its nature.

People in a cultivation culture are idealists, particularly in their approach to people. They are prone toward looking at things in a holistic or assimilative manner. They seek ideal solutions, are possibility oriented, and stay focused on values. They are very aware of how they feel about things; emotions are used to understand what is going on around them and to give them signals about the right thing to do. Ethics are important. They dedicate themselves to causes. Cultivation culture people pay close attention to process, believing that if they nurture the process, valuable results will come about. By nature, they tend to be receptive, and they have an easy time making commitments. Their approach to life is optimistic and positive.

LEADERSHIP, AUTHORITY AND
DECISION MAKING

Leaders in the cultivation culture are catalysts who stimulate the bringing about of a desired process or goal. They do this by creating the potential for growth among the people working in the organization. Cultivation culture leaders are expanders and enlargers of their people and organization. When people, ideas, products, and services develop and mature, the leaders see that effective harvesting occurs.

Leaders in this culture also must envision possibilities. They strive to get their people dedicated to a common vision. This often entails a lot of work developing and articulating purposes and visions for people and, having done that, building and keeping them committed.

Leaders have great faith and optimism in the integrity and untapped talents of their people. Because they trust their people, leaders can concentrate on stewardship and unleashing the potential of their people. They are good at getting people to believe that they can lick the world in the morning and have the rest of the day off. The death of Bill Bernbach in 1982 brought a tremendous outpouring of emotion from the entire advertising community. One letter testified to the faith Bernbach had in people. Sam Kurtz, who at one time worked for DDB, wrote:

> Nobody mentioned, for example, that [Bernbach] hired an awful lot of bums—neurotics, drunks, complainers, kooks of every description . . . because he saw a vein of talent in them . . . I personally and literally owe him my life. He brought out what talent I had . . . You know what his chief deep-down achievement was? Let me tell you. A lot of those people didn't have homes. They had a place to live, but they didn't like going there. So he made home for them at DDB . . . I used to see them playing ball in the corridors. "Why don't you go home?" I used to ask. One would say he was waiting for his wife to put the kids to bed. Another would say the trains were too crowded. A third would say that he was waiting for type. And on and on. But the truth was that they hated to leave the office, they were so happy and comfortable there.[17]

Leaders in this culture concentrate on creating conditions for success. People are empowered to think independently and to let their creative juices flow. Trust and commitment provide the glue and the order. Leaders believe that people will dedicate themselves to doing their best if they are committed to the purposes of the organization and are given an opportunity to grow and develop. Each person's contribution to the attainment of overall objectives is important. The organization's success is assured if people perform well in their individual jobs. Accordingly, leaders devote a great deal of time putting all the resources in place so their people have what they need to realize their own—and the organization's—potential. Leaders also give people considerable individual attention. Mo Siegel, founder of Celestial Seasonings, said, "What is important is creating a condition in which the work force feels better about their lives, because what is good for labor should be good for management and vice versa."[18]

People are strongly encouraged to think and act creatively and to take reasonable risks. Leaders advocate discovery and experimentation, and welcome fresh ideas. They allow eccentricity and find the offbeat intriguing. They ask people to challenge the status quo and question dogma and tradition.

The cultivation culture values continuous improvement and is attentive to the future and new possibilities. What may seem initially to be impractical or even unsound is given full consideration and exploration. They relish the sea of change. Progress, small and large, makes things tick. Leaders work at uplifting their people, helping them to be more complete and to fulfill their aspirations. In its early years, Apple Computer, Inc., the computer manufacturer, saw itself as manufacturing and selling a "revolutionary tool" that unleashes computing capability for individuals. Until Apple developed the personal computer, the people at Apple knew that the power of computing was reserved for large companies or governments. The Apple leadership was determined to make work on the personal computer an adventure. People with dreams would make a difference. This meant that managers had to take the right attitude toward employees, who in turn had to trust the motives and integrity of their supervisors. At Apple, management was responsible for creating a "productive environment where Apple values flourish . . . Apple wants managers who are expanders, not controllers of people."[19]

More often than not, leaders in a cultivation culture are enthusiastic, positive, and insightful; they strive to energize and stimulate their people. Individual and organizational motivation is at the top of their agenda. Cultivation culture leaders are often activists who provide a great deal of forward thrust for their organizations; they keep their people psyched up, enthused, and involved. The organizations are lively, spontaneous, and magnetic. People who work for these leaders feel like there is never a dull moment. Work is punctuated with frequent periods of high intensity. "Why not shoot for the stars?" is the slogan.

Leaders frequently speak for their people. They emphasize values and putting these values into operation. People follow in this culture because they believe what the leadership and the organization espouse; because they value having opportunities for growth, self-expression, and the fulfillment of potential; and because they like feeling special.

Like the other core cultures, the cultivation culture has its own brand of power. In this culture, *charismatic* power operates. *Charisma* is the quality of leadership that captures the imagination and inspires allegiance and devotion. People in a cultivation culture have power by "inspiring" and "capturing the imagination" of others. Leaders are typically the most inspiring people in this culture. They infuse life into the organization. Their ability to motivate others to grow forms the base of their power. At the same time the ability to capture the imagination builds commitment, trust, and faith. In essence, people are influenced in this culture by being enlivened by being brought more life.

Decision making is typically dynamic and longitudinal. The culture is in motion a great deal. There is a natural instinct to look for different ways to do things, and numerous possibilities, hunches, or alternatives are generated. Experimentation is frequent, particularly if it is action oriented. Testing and retesting goes on much of the time. The atmosphere is often highly collegial and interactive; people build on one another's ideas; participation is high. At the same time, however, individuals are free to branch out in independent directions. Personal values often drive what gets investigated, pursued, and eventually decided upon. Importantly, the purposes of the organization provide the final framework for deciding what will be implemented or brought to customers or constituents.

People proceed rapidly, following their instincts, aspirations, and inspirations. People usually have multiple foci which incorporate as much as is deemed relevant. Decisions are made quickly, but can be changed just as easily in this fluid culture. They are always open to modification by new information, and there is always room for development, improvement, and expansion. Life is more open-ended than in any other culture. People often think that "decision making" is something of a misnomer because things are so fluid and continuous. Reaching a decision is more like pausing along the way than anything else.

STRUCTURE AND RELATIONSHIPS

Despite what the Table of Organization of a cultivation culture might look like on paper, its true structure is circular or wheel-like. People are free and encouraged to interact with everyone else in the organization. Most activities and functions are decentralized, with minimal lines of authority. Individually or collectively, people connect with one another to create something, to pursue an area of investigation, to experiment, and to develop a product. There are few rules, policies, or procedures. People are trusted to do what is in the best interests of the organization and its customer and constituent base.

Bill Gore, president of W. L. Gore & Associates, a fabric manufacturer in Delaware, designed an organizational system called the "lattice organization." In this system, each person in a "lattice" can interact directly with every other person. According to Gore, this organization has several attributes:

- No fixed or assigned authority.
- Sponsors, not bosses.
- Natural leadership defined by followership.
- Person-to-person communication.
- Objectives set by those who must make them happen.
- Tasks and functions organized through commitments.

Gore's company is without titles, hierarchy, or any of the conventional structures associated with enterprises of its size. *Inc.* magazine calls Gore's approach a "system of un-management."[20]

Relationships in this culture are free flowing, flexible, and built on trust and commitment. The presence of good will is assumed. Relationships among people are viable and continuous because of "covenants" with one another. Roles per se are much less prevalent than person-to-person connections. People get together with others to accomplish something or they operate on their own; either modality is OK. Accordingly, people have considerable autonomy. In some ways, things just happen in the cultivation culture.

Relationships change a lot. Life is fast-paced. People may interact freely in making necessary contacts without fear of exceeding someone's authority or station. This culture values adapting from the old to the new. The company structure at 3M is more "biological" than that found on a traditional organizational chart. Fluid movement is a hallmark within the company. New divisions are created when new products develop sufficiently large sales. Inventors are free to sell their ideas anywhere within the company and are not limited to their own divisions. When a potential product is adopted, the inventor moves in with the new divisional sponsor. They are also free to employ marketing professionals from other parts of 3M to work on the new project.[21]

Information flows freely and little is kept secret. Internal competition is minimal because people help one another out a great deal. Mutual encouragement prevails. The humanistic side to issues—how people feel about things—is important in this culture because information is derived from feelings.

You name it—generalists, specialists, functionalists—they all are bred in a cultivation culture. Like the collaboration culture, diversity is highly valued. However, in a cultivation culture diversity is utilized more for pure creativity than for carefully designed team functioning. In this culture, individual talent can grow and move in a rather wide range of directions. This is less true of a collaboration culture.

Teams are built for a particular task or project and are then disbanded. Individuals may then work on their own for a long time. It all depends. Things are more unpredictable in this culture than in any other.

Commitment is very high in a cultivation culture. You have read how it is a key element in the operation of this culture. People develop strong commitments to one another, to the organization, and to the purposes of the organization. Accordingly, there is a resolute quality to relationships that makes it difficult to give up on people because to do so is to break commitments.

The climate—what it feels like in a cultivation culture—is typically relaxed, spirited, and energetic. Optimism and positive attitudes tend to prevail. A feeling of freedom permeates the organization. People behave as if they have few things to worry about. Communications are open and direct. There is often a feeling of calmness present.

STAFFING AND PERFORMANCE MANAGEMENT

In many cultivation cultures, employees do the recruiting. Word-of-mouth brings people in or employees know of someone whom they are convinced will fit into the organization.

Entry into a cultivation culture often hinges on faith, commitment, and values. The key question concerning a new recruit is: "Will this person be committed to our beliefs? Does he or she value the purposes of our organization?" John Schaeffer, founder of Real Goods Trading Corporation, an alternative-energy products company in Ukiah, California, says that his company's employees "can sell things better if they believe in [our] crusade for energy conservation." In discussing his decision to hire an individual as operations manager, Schaeffer highlighted the candidate's level of commitment that came through in a series of three interviews: "He said his heart and soul lay in getting the environment on a cleaner track. It became clear that he fit."[22]

Talent and capability issues are important, but because the culture is so oriented to growth and development, these concerns are less critical on the front end. These organizations work to fit individuals into their workplace and then develop them. Talent and capability are cultivated after people have joined the organization. In addition, the cultivation culture assumes and believes that individuals have unique talents that need to be developed and broadened. Therefore, the recruit's personal belief system is the key issue. If there is a mismatch on this issue, the probability of subsequent failure goes way up.

Managing performance is essentially a matter of cultivating, developing, and guiding people's commitment, growth, and production. It works because the central rationale in cultivation cultures is that people are good-willed, that every person has unique talents that can be nurtured and focused on valuable goals, and that people within the organization believe in the organization. Considerable emphasis is placed on clarifying the purposes of the organization, articulating its vision and mission, and creating the conditions that best promote that vision.

The organization gives considerable leeway and freedom to make a contribution. Jobs are matched to people, not people to the jobs. People with many talents are encouraged to do many things. The work generally offers considerable variety, novelty, and challenge. Eccentricity and mistakes are tolerated; within reason and capability, people are given as many chances to succeed as possible. This culture gives people a great deal of individual attention. The culture's expectations of an individual soon become tailored to that specific person. Bosses "weed out" the garden—they remove barriers to self-expression, growth, and production. "Promotion" and "demotion" are a misnomer because they connote a hierarchical system, which the cultivation culture is not. Instead of promotion or demotion, people take on more responsibility or narrow the scope of their responsibility.

Feedback is typically generous and gratefully received. However, it focuses more on acknowledging what a person has said or done and builds upon that. This differs from the feedback process in the competence culture, which measures the extent of a person's achievement.

Training and education are prevalent, but the focus is different than in other core cultures. In the cultivation culture, the emphasis is on the discovery and development of a person's unique talents. When people are challenged—and this occurs often—they are challenged to perform to the limit of their potential. The belief is that as people grow, the organization grows. People are mentored and sponsored; they educate one another. On-the-job training is customary.

The culture encourages initiative and prizes individual contribution. Job rotation is frequent. People have every opportunity to stand out in their own right and be recognized as unique. W. L.

Gore & Associates warns prospective employees by letter what to expect: Bill Gore says, "If you are a person who needs to be told what to do and how to do it, you will have trouble adjusting . . . An associate has to find his own place to a high degree. There's no job description, no slot to fit yourself into. You have to learn what you can do."[23] Jack Dougherty discovered that the letter from W. L. Gore was more than cautionary. Arriving for his first day of work in 1976, Bill Gore shook his hand and said, "Why don't you look around and find something you'd like to do." Dougherty, then a 23-year-old business school graduate, wandered around to various plants in the company for several weeks until he found himself interested in turning Gore-tex into fabrics for parkas and camping gear. Once committed, Dougherty put on his jeans to work on the production line. By 1982, Dougherty was responsible for Gore-tex's advertising and marketing.[24]

As you know, the cultivation culture runs on trust and commitment. Nowhere is this more evident than in performance management. The organization trusts people to want to do their very best, and it takes for granted their commitment to the purposes of the organization. Because of this, the culture can do what it does. Without that trust and commitment, all else would falter. Performance management is mostly placing people in the organization where they can make the greatest contribution.

Ill will gets a person in hot water. If someone is not trusted or is perceived to be uncommitted, it is more than likely that person will be asked to leave. Ineffective performance or lack of capability will only cause a person to be "re-placed" in a more suitable place within the organization. This will be interpreted as a mismatch that can be put right with an opportunity to find the right soil for this person to grow and develop. On the other hand, a person who cannot be trusted or is accused of self-aggrandizement is not going to make it in this culture.

Process gets more emphasis than results in a cultivation culture. If you dedicate yourself and the organization to the process of what you are doing, the desired results will naturally accrue. One necessarily follows from the other. A good job of tilling your soil, planting your seeds, watering, and harvesting will yield the results you seek because you had faith and let nature take its course.

Conflict itself is seen as valuable and potentially enlightening. However, it is encouraged only in a muted fashion. Yelling and screaming are not OK; the quiet, calm, and reasonable exploration of differences is.

The people who fit with the greatest ease in the cultivation culture are intuitive and idealistic. They treasure their own self-expression and independence and they are open to change and development. What is important to them are possibilities and the future, their values and beliefs. They tend to be creative. Virtuosos fit well in this culture. Being part of something bigger than themselves is important. They like to help others grow and reach their potential.

WHERE DOES A CULTIVATION CULTURE FIT?

The cultivation culture is more naturally suited for enterprises that emphasize raising the human spirit and furthering the realization of values. Helping to make people and the world more perfect best characterizes it. The cultivation culture touches the emotions and the heartstrings of people much more so than any of the other three cultures.

Enterprises that fit logically into this cultural framework are characterized by purposiveness and will power and the striving toward the realization of ideals. Examples include artistic enterprises, religious enterprises, or any product or service designed to accomplish purposes of a higher order for its customers.

Appendix A of Chapter Six
STRENGTHS OF THE CULTIVATION CULTURE

It does a good job of building commitment and dedication among its people.
It makes people feel cared for, nurtured, and special.

It offers considerable opportunities for growth, development, and the realization of potential.

It highly values people's aspirations and hopes.

It values creativity.

When effective, trust is abundant and people feel accepted for who they are and what they can become.

Resolution and determination come naturally.

It values and nourishes differences and diversity. When successful, people feel fulfilled and gratified that they are contributing to worthwhile goals.

When effective, individual talent is more fully tapped and utilized than in the other three cultures.

It is amenable to change and adaptation.

It is naturally inclined to be socially responsible.

People feel inspired and enlivened.

It strongly encourages self-expression. Empowerment is natural. People have much freedom and autonomy. Virtuosos, eccentrics, and individualists can do well.

It places a high value on training and education.

People can make mistakes and not be punished.

When successful, optimism and positivism prevail.

It can make great, worthwhile contributions to people, communities, and society. More than in any other culture, higher order values are realized and put into action.

Appendix B of Chapter Six
WEAKNESSES OF THE CULTIVATION CULTURE

In excess, it lacks direction and focus. Too much self-expression, left unchecked, takes the organization down a fair number of primrose paths. The organization goes in so many directions at once that it squanders its energies.

In excess, many things do not get finished, and projects lay on the shelf.

Taken to extremes, people become moralistic and overly judgmental.

It is prone toward sweeping problems under the rug, particularly problems between people.

It is prone toward having a hard time with the coordination of people and activities.

It is prone toward playing favorites.

It is prone toward ineffectiveness in highly competitive situations, because it is so committed to good will. It has a hard time knowing how to respond to competitors who don't play fair.

People try too hard for change and newness, pushing for change when it is more effective to keep things the way they are.

"Hard" data (particularly systems and quantitative data) may get screened out.

People make too many delays because they face too many choices.

People try too hard for perfect solutions. The culture is prone toward too much idealism.

It is prone toward inefficiency.

Details, particularly mundane ones, are easily overlooked.

In excess, people are overly sentimental.

People let intention prevail to too great an extreme, resulting in a minimization of performance and results. In extreme, the organization could be slow to realize it is in trouble.

People feel like they never arrive or make real, and lasting contributions. This occurs because the culture is so possibility oriented and there is always a tomorrow to finish things.

People get burned out and overburdened because the culture can become so compelling.

People are too oriented against controls, resulting in inattentiveness to areas or issues that need control.

Some ideas or initiatives outlive their usefulness but are not discontinued.

NOTES

1. R. Levering, M. Moskowitz, and M. Katz, *The 100 Best Companies to Work For In America* (Reading, MA: Addison-Wesley, 1984), p. 43.

2. Ibid., pp. 218–19.

3. A. H. Maslow, *Toward A Psychology of Being* (Princeton, NJ: D. Van Nostrand, 1968).

4. S. R. Maddi, *Personality Theories: A Comparative Analysis* (Homewood, IL: Dorsey Press, 1968), pp. 83, 278.

5. M. Scott and H. Rothman, *Companies With a Conscience: Intimate Portraits of Twelve Firms That Make a Difference* (New York: Carol Publishing Group, 1992), p. 64.

6. Ibid., p. 64.

7. Levering, Moskowitz, and Katz (1984), p. 5.

8. Ibid., p. 2.

9. Ibid., p. 48.

10. Quoted in D. L. Kirp and D. S. Rice, "Fast-Forward—Styles of California Management," *Harvard Business Review*, 1 (January–February 1988), p. 80.

11. Levering, Moskowitz, and Katz (1984), p. 50.

12. Ibid., p. 49.

13. Ibid., p. 88.

14. Ibid., p. 221.

15. Ibid., pp. 221–22.

16. Ibid., pp. 41–42.

17. Ibid., pp. 90–91.

18. Ibid., p. 41.

19. Ibid., pp. 12–13.

20. Ibid., pp. 128–29.

21. Ibid., pp. 222–23.

22. B. Bowers, "Follow the Leader: Recruiting Employees Who Share the Corporate Vision Requires a Delicate Balancing Act," *Wall Street Journal Reports* (November 22, 1991), R16.

23. Levering, Moskowitz, and Katz (1984), p. 128.

24. Ibid., p. 129.

Chapter Seven

The Genesis of Organizational Culture

The reader is encouraged to look at the series of five summary tables on the four core cultures. Each table focuses on an important dimension as it relates to each culture.

These are the

Strategic focus

- Definition of success
- Way to success
- Approach with customer/constituents

Leadership and management focus

- Leadership focus
- Management style

Structural focus

- Organizational form
- Role of the employee
- Task focus

Power focus

- Nature of power/authority
- Approach to decision making
- Approach to managing change

Relationship focus

- Key norms
- Climate

TABLE 7–1
The Four Core Cultures (Strategic Focus)

Definition of Success	Way to Success	Approach w/Customers and Constituents
CONTROL		
Dominance The Biggest	Get and keep control	Controlling Only game in town
COLLABORATION		
Synergy	Put a collection of people together, build them into a team, engender their positive relationships with one another, and charge them with fully utilizing one another as resources.	Partnering We did it together
COMPETENCE		
Superiority The Best	Create an organization that has the highest level of capability and competence possible and capitalize on that competence. Pursue excellence.	Offer the most superior value Nothing else compares One of a kind State of the art
CULTIVATION		
Fuller realization of Potential Growth	Create and provide the conditions whereby the people within the organization can grow, develop, and strive to accomplish the highest order purposes possible.	Realize possibility and potential more fully Fulfillment Uplift/enrich

The tables will be particularly helpful as you apply the concepts of the four core cultures to organizational and management development efforts.

Appendix A is a summary of the work of others which corroborates the concepts of the four core cultures explored in this book. While no one has proposed the existence of *core* organizational cultures, a sizable number of experts from varying fields have

TABLE 7–2
The Four Core Cultures (Leadership and Management Focus)

Leadership Focus	Management Style
CONTROL	
Authoritative/directive	Methodical
Maintain power	Systematic
Conservative/cautious	Careful
Commanding	Conservative
Firm/assertive	Policy and procedure oriented
Definitive	Task driven
Tough-minded	Impersonal
Shot caller	Prescriptive
Realistic	Objective
COLLABORATION	
Team builder	Participative
Coach	Collegial
First among equals	Democratic
Participative	Relational
Representative	Supportive
Integrator	People Driven
Trust builder	Personal
Commitment builder	Emotional
Realistic	Adaptive
Ensure utilization of diversity	Informal
Bring in the right mix of talent	Trusting
COMPETENCE	
Standard Setter	Task driven
Architect of Systems	Objective
Visionary	Rational/analytical
Taskmaster	Intense
Assertive, convincing persuader	Challenging
Set exacting expectations	Efficient/crisp
Stretch people/push limits	Impersonal
Recruit the most competent	MBO/MBR in nature
Incentivizer	Hard to satisfy
Emphasize what's possible	Formal
Challenge subordinates	Emotionless

TABLE 7–2
(Concluded)

Leadership Focus	Management Style
CULTIVATION	
Catalyst	People driven
Cultivator/harvester	Personal
Empower/enable people	Relaxed
Potentializer	Emotional
Commitment builder	Attentive
Inspire/enliven people	Promotive
Promoter/motivator	Nurturant
Steward	Humanistic
Maker of meaning	Enabling/empowering
Appeal to common vision	Purposive
Foster self-expression	Committed

proposed the existence of organizational forms that strongly support the reality of four essential organizational cultures.[1]

ORGANIZATIONAL CONTENT AND PROCESS

When looked at together, the four cultures reveal a number of underlying patterns (see Exhibit 7–1 on page 112). The fundamental, underlying pattern is illustrated by two basic axes that, when combined with one another along two separate axes, yield a four-element table. Each section in this four-element table represents one of the four core cultures.

The vertical axis considers what an organization pays attention to, or the *content*. The horizontal axis considers how an organization makes decisions, forms judgments, or the *process*.

The content axis is bounded by *actuality* and *possibility;* the process axis is bounded by *impersonal* and *personal*.[2]

Exhibit 7–1 shows that collaboration and control are actuality cultures, cultivation and competence are possibility cultures, collaboration and cultivation are personal cultures; and control and

TABLE 7–3
The Four Core Cultures **(Structural Focus)**

Organizational Form	Role of the Employee	Task Focus
CONTROL		
Hierarchy	Compliance Adhere to role requirements Functionalist Actualizer Serve the functional pursuits of the organization Be realistic Be useful Follow directives	Functionalist Individuals stay within a function Specialties channeled into service of functions
COLLABORATION		
Group Cluster	Collaborate. Be a team player Contribute to the whole effort Utilize others as resources Actualizer Honor diversity Commitment to the whole organization Identify with the whole organization Take initiative Generalist	Generalist Individuals serve in numerous functions .
COMPETENCE		
Matrix Adhocracy	Be the best at what you do Be an expert Advance knowledge Be creative Be a possibilizer Serve the conceptual and theoretical pursuits of the organization Function independently Demonstrate competence Specialist	Specialist Individuals stay in technical specialty Functions channeled into service of specialties

TABLE 7–3
(Concluded)

Organizational Form	Role of the Employee	Task Focus
CULTIVATION		
Wheel-like Circular Lattice	Be creative Express yourself Be willing to change, develop, grow Be a possibilizer Be all you can be. Realize your potential Cooperate Believe in what the organization believes in Be committed Be versatile	All three above Individuals do all three

competence are impersonal cultures. Each core culture is a unique blend of one content element and one process element. Control is an *actuality–impersonal* culture; competence is a *possibility–impersonal* culture; cultivation is a *possibility–personal* culture; and collaboration is an *actuality–personal* culture.

It is important to preface the detailed discussion of these matters by noting that the qualities and characteristics associated with the content and process axes are organizational and cultural preferences or central tendencies. They are not exclusionary; having a preference for one does not preclude involvement in the other. For example, you will learn that facts are a content area of an actuality organization. This means that an actuality organization has a *preference for* focusing on facts. It does not mean that facts are all that an actuality organization deals with or that a possibility organization never attends to facts. These organizational preferences are like right- and left-handedness in people; they exhibit a preference for one or the other, but they also use both. In organizations, one simply predominates or is more central to how things work. Readers will also find it helpful to refer to summary

TABLE 7–4
The Four Core Cultures (Power Focus)

Nature of Power/Authority	Approach to Decision Making	Approach to Managing Change
CONTROL		
Role/position Titular	Very thorough Methodical Detached Impersonal Push for optimal solution Push for certainty Prescriptive Cause to effect Formula oriented Very objective	Mandate it Resistance to change
COLLABORATION		
Relationship	Participative Collegial Democratic Consensus oriented Emotional Experimental. Lots of brainstorming Personal/interpersonal People driven Organic/evolutionary Subjective Trusting	Team calls for change Open to change
COMPETENCE		
Expertise	Very analytical Objective Scientific Detached Impersonal Efficient Rational Principle and law oriented Formula oriented Formal logic Emotionless	Achievement goals drive change Open to change

TABLE 7–4
(Concluded)

Nature of Power/Authority	Approach to Decision Making	Approach to Managing Change
CULTIVATION		
Charisma	Purposive People driven Organic/evolutionary Very subjective Emotional Committed Participative Dynamic Personal/interpersonal Adaptive Informal	Embrace/assume change Change is automatic

Tables 7–1 to 7–5 to refresh their thinking about the four core cultures during the discussion of the four-element Organizational Content and Process exhibit and its dimensions.

ORGANIZATIONAL CONTENT: WHAT THE ORGANIZATION PAYS ATTENTION TO

At the most fundamental level, every organization focuses either on what is actual or what is possible. Actuality has to do with what is; possibility has to do with what might be.

The content of an actuality culture is:

- Concrete, tangible reality.
- Facts.
- What has occurred in the past and is occurring in the present.
- Actual experience/actual occurrence.
- What can be seen, heard, touched, weighed, or measured.
- Practicality/utility.

TABLE 7–5
The Four Core Cultures (*Relationship Focus*)

Key Norms	Climate
CONTROL	
Certainty	Serious
Systematism	Realistic
Objectivity	Matter-of-fact
Order	No nonsense
Stability	Low-key
Standardization	Air of secrecy
Utility	Restrained
Realism	Formal
Discipline	Impersonal
Predictability	Steady, regulated pace
Accumulation	Unemotional
COLLABORATION	
Synergy	Harmonious
Egalitarianism	Personal
Diversity	Spontaneous
United we stand, divided we fall.	Work hard/play hard
People interaction	Esprit de corps/camaraderie
Involvement	Can do
Harmony	Busy pace
Realism	Emotional
Pragmatism	Open/free and easy
Complementarity	Lively give-and-take
Spontaneity	Trusting
COMPETENCE	
Professionalism	Competitive
Meritocracy	Intense pace
Pursue excellence	Unemotional
Creativity	Impersonal
Continuous improvement	Serious
Competition for its own sake	Professional
Craftsmanship	Rigorous
Don't rest on your laurels	Prideful
Efficiency	Spartan
Autonomy/individual freedom	Rational
Accuracy	Formal

TABLE 7–5
(*Concluded*)

Key Norms	Climate
CULTIVATION	
Growth and development	. Lively/magnetic
Humanism	Committed
Faith	Spirited
Commitment/dedication	Personal
Involvement	Emotional
Creativity	Cooperative
Purpose	Relaxed and fast paced
Let things evolve	Open
Freedom to make mistakes	Informal
Shoot for the stars	Compassionate
Subjectivity	Caring
Values are paramount	Giving/generous

The content of a possibility culture is:

- Insights.
- Imagined alternatives.
- What might occur in the future.
- Ideals/beliefs.
- Aspirations/inspirations.
- Novelty.
- Innovations/creative options.
- Theoretical concepts or frameworks.
- Underlying meanings or relationships.

The actuality organization lives in actual experience and reality. The possibility organization lives in anticipation.

Compare our core culture pairs—control and collaboration, the actuality cultures, are focused on getting today's job done and the materials their customers or constituents bring them. You control what *is*, much less what might be. The control culture pushes for certainty and realism. There is a predominant empha-

EXHIBIT 7–1
Organizational Content and Process

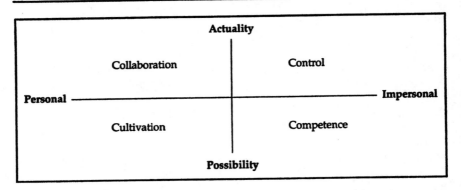

sis on facts and factual material. The more predictability, the better. You collaborate with real people; you can't collaborate with an ideal.

The collaboration culture focuses on what is going on today within the organization and with its customers and constituents. The future will come about by working emphatically on today's real concerns with team members, customers, and constituents. Synergy is a tangible, present tense, experience. Both control and collaboration cultures are highly pragmatic. People in both cultures have a natural affinity for the tangible and focus primarily on real, daily issues and concerns.

By contrast, competence and cultivation, the possibility cultures, are focused on getting tomorrow's job done and the achievements or potentials to be accomplished. The competence culture strives for superiority and being the best; it sets higher and higher standards. The focus is on continued achievement and the ongoing pursuit of excellence. The cultivation culture emphasizes inspiration, fulfillment, and growth. Both cultures are centered on ideals and imagined alternatives, are creative by nature, and are engaged in explorations beyond past and present realities. For competence and cultivation life is beyond the horizon. Both are more vision oriented and largely focused on reaching higher level purposes.

Control and collaboration emphasize doing and producing now. Competence and cultivation emphasize reaching toward the future.

ORGANIZATIONAL PROCESS: HOW THE ORGANIZATION DECIDES

The other fundamental dimension that underlies organizational culture is how the organization forms judgments. Every organization emphasizes either impersonal analysis or personal/interpersonal involvement. Impersonal analysis entails the use of detached reasoning. Personal and interpersonal involvement entails the use of people actively judging, deciding, and acting.

The process of an impersonal culture is:

- Detached.
- System, policy, and procedure oriented.
- Formula oriented.
- Scientific.
- Objective.
- Principle and law oriented.
- Formal.
- Emotionless.
- Prescriptive.

The process of a personal culture is:

- People driven.
- Organic/evolutionary/dynamic.
- Participative.
- Subjective.
- Informal.
- Open-ended.
- Important-to-people oriented.
- Emotional.

An impersonal process organization reaches decisions by relying on systems, formulas, and methods that are external to the minds and hearts of its people. A personal process organization reaches decisions by unleashing what is internal to the minds and hearts of its people.

Look again at our core culture pairs. Control and competence, impersonal cultures, go about decision making by prescribing systems, formulas, and methods for people to follow. In the control culture, people are provided with the way to get and keep control and to accomplish dominance. It emphasizes precedents and requires objectivity and order. It formulates and establishes policies and procedures. People are directed in the control culture.

The competence culture establishes achievement goals and deploys people to reach them. It sets standards and requires people to adhere to them. The scientific method is often used. The culture relies on formal logic and emphasizes the empirical approach.

Both cultures, by nature, tend to be regulated and push for predictability. System development and implementation is very important. Control and competence are role driven, meaning that the organization establishes the roles it requires and then fits people into those roles. It values status, rank, and issues of formal organizational life highly. The two kinds of power—role power and expertise power—associated with control and competence are impersonal by nature. They spring from outside people and are externally verifiable. Both cultures implicitly require their people to place their trust in the organization.

Life is different, however, in collaboration and cultivation cultures, both of which are personal cultures. There judgments and decisions are made by empowerment. In these cultures, the organization places its trust in its people. Management not only allows mistakes, but anticipates them. The collaboration culture is driven by synergy and by people teaming up and utilizing one another as resources. Brainstorming and collective experimenting go on all the time. The cultivation culture emphasizes the fulfillment of people's potential and aspirations. In both cultures, decisions and actions are determined by the people working either individually or collectively. The action taken by these two cultures is much more organic and evolutionary. The spontaneous judgment and wisdom of people carries the organization forward. Contrary to the two impersonal cultures, status and rank are of minimal importance in the collaboration and cultivation cultures. The power associated with collaboration and cultivation cultures—relationship power and charisma power—are personal by nature and spring from within people. They are verifiable only by relying upon people's experiences with one another. Because the collaboration and cultivation cultures are people driven, organizational roles are more often than not fitted to the people.

Control and competence cultures place impersonal principles and formulas first; they are what move the people and organization ahead. Collaboration and cultivation cultures place people capability, spirit, and motivation first; they move the organization ahead.

THE FOUNDATION OF EACH CORE CULTURE

The foundation of each core culture rests on what each culture focuses on and how each makes decisions. Said another way, each core culture is uniquely defined by the kind of input that is important to it and by the process it relies on to form judgments and make decisions.

The Control Culture

The control culture is an actuality–impersonal culture. What it pays attention to most is concrete, tangible reality; actual experience; and matters of practicality and utility. Its decision-making process is analytically detached, formula oriented, and prescriptive.

If you could magically transform a control culture into a working person, the best representation would be a surgeon. A surgeon has to get control, to stay in command of the situation, and to be very thorough and methodical. Surgeons must stay attentive to the realities of their task, and they must be analytically detached and impersonal when deciding what to do.

There is much that is no-nonsense and matter-of-fact about the control culture. It is a serious and highly realistic world. Not much speculation goes on. Discipline and definitiveness are naturally strong elements. Considerable emphasis is placed on objective certainty and on having information and methods that can be relied upon. Predictability is very important.

Given its content and process nature, this culture is highly task oriented, particularly concerning daily tasks. Getting the job done on a consistent, regular, and current basis is very important. Strong emphasis is placed on rules and adherence to them. While production is clearly important for every culture, it holds center stage in a control culture.

In general, something becomes known or believed in this culture when organizational system goals and reality coincide. Systematism counts—the organization as a system comes first. Accordingly, the design and framework for information and knowledge in the control culture is built essentially around the goals of the organization, and the extent to which those goals are met.

Objectivity lies at the very center of the control culture. How decisions get made hinges on content and methods that exist outside people. In the control culture reality lies outside people. As a result, people are used essentially in the service of externals: what gets worked on and how the work is done.

Emerson Electric Company of St. Louis—by all accounts, a consistently successful business organization—is an excellent example of a control culture. Charles F. Knight, chairman and CEO, describes how Emerson does things in order to succeed.

> The driving force behind [Emerson's continued success] is a simple management process that emphasizes setting targets, planning carefully, and following up closely. The process is supported by a long-standing history of continuous cost reduction and open communication and is fueled by annual dynamic planning and control cycles. Finally, it is nourished by strongly reinforced cultural values and an approach to organizational planning that is as rigorous as our approach to business planning . . .
>
> [M]ore than half of my time each year is blocked out strictly for planning . . . from November through June, selected corporate officers . . . and I meet with the management of every division for a one- or two-day planning conference . . . The mood is confrontational— by design. Though we're not trying to put anyone on the spot, we do want to challenge assumptions and conventional thinking and give ample time to every significant issue. We want proof that a division is stretching to reach its goals, and we want to see the details of the ac-

tions [that] division management believes will yield improved results . . . A division president who comes to a planning conference poorly prepared has made a serious mistake . . .

The measure of Emerson managers is whether they achieve what they say they will in a planning conference. We track the implementation of our plans through a tight control system. That system starts at the top, with a corporate board of directors that meets regularly and plays an active role in overseeing our business.[3]

Knight's description of how quality of communication is measured at Emerson illustrates the importance of production in a control culture: "We claim that every employee can answer four essential questions about his or her job: 1. What cost reduction are you currently working on? 2. Who is the 'enemy' (who is the competition)? 3. Have you met with your management in the past six months? 4. Do you understand the economics of your job?"[4] Emerson's principal focus is clearly on actuality and its decision-making process is almost exclusively impersonal.

The Collaboration Culture

The collaboration culture is an actuality–personal culture. Like the control culture, it pays a great deal of attention to concrete, tangible reality; actual experience; and matters of practicality and utility. However, its decision-making process is people driven, organic, and informal.

While the surgeon best represents the control culture, a nurse best embodies the working person counterpart in a collaboration culture. Nurses are focused on the immediate, real, and physical matters that pertain to their patients. Serving the sick, infirm, or injured has immediacy. Patients need to be treated now, not down the road. In addition, a nurse typically works as part of a medical team. Judgments about what treatment is needed and how to conduct it usually come from the team. The process is dynamic and highly participative. The relationship between the patient and the nurse is extremely important. What the two accomplish together—the synergy created—contributes considerably to the effectiveness of the outcome.

Synergy is the natural definition of success in the collaboration culture (see Chapter 4). Synergy itself captures the content and process of this culture. When you take two people or chemicals and combine them, and when you get 5 instead of 4, you have combined two *actuals*, two *realities*, by utilizing an *organic* or *dynamic* process. The content at issue here is not what might be or theory. It *is*. It is tangible. And, the process at issue here is not detached, it is *involved*. The dynamic process enables people to empower one another and deliver what is within each to the other in order to bring about something more. The process is highly biological.

Interaction and involvement are central elements. Little happens that is solitary or solely independent. Life is busy and spontaneous. Harmony and cooperation are essential elements in this "can do" culture. The process is inherently win-win.

This culture probably puts more trust in its people than any of the other three. Because its content is immediate reality and actual experience, and its process is driven by connections between people, the risk of the organization being surprised by problems is higher in this culture than in any other. While the cultivation culture is also people driven, its focus on the future and on possibilities gives it more time to deal with unforeseen events. The collaboration culture must be more adaptive, ready, and able to make adjustments, than the other three cultures.

The nature of knowledge in a control culture is tied to the connection between system goals and reality; in a collaboration culture, the connection is between people's experience and reality. The organization moves ahead through the collective experience of people from inside and outside the organization. Collaboration culture people know something when collective experience has been fully utilized.

Chaparral Steel in Midlothian, Texas, is an example of a collaboration culture.

> Nine hundred thirty people work at Chaparral Steel, and there are 930 different rates of pay. No two people are paid the same here because Chaparral has created an unusual way of paying workers, based on their job, seniority, versatility, skills, and training needs. The system may drive the payroll department crazy, but the employees love it.[5]

Gordon Forward, Chaparral's president and CEO, says: "We try to treat employees as adults. What we're doing is based on trust. Docking pay . . . [is] a distrust system. Time clocks also show distrust. They say the company doesn't trust you at 8:00 in the morning, and they still don't trust you at 4:00 in the afternoon."[6] Chaparral Steel doesn't favor employee suggestion programs because they undermine teamwork and work counter to brainstorming. Instead of paying for suggestions, the company shares 8.5 percent of its gross profits with its employees. All employees share in the rewards of increased productivity, but only of course when there are profits.[7]

The Competence Culture

The competence culture is a possibility–impersonal culture. It pays most attention to potentiality, imagined alternatives, creative options, and theoretical concepts. Its decision-making process is analytically detached, formula oriented, scientific, and prescriptive.

The best occupational counterpart to the competence culture is the research scientist. A research scientist focuses on what might be and looks at data, information, patterns, and underlying meanings and relationships. Anticipated outcomes or hypotheses drive the scientist's work. He or she investigates insights. The essence of science itself is to go beyond what we already know or have learned, and to determine what else is possible. Theoretical development is often at stake. The scientist's process is the scientific method. This means that the scientist must stay analytically detached, objective, and follow prescribed formulas for making judgments. Possibility is determined objectively by the relevant scientific principles.

This is a world of external verification and merit in which objective feedback is probably more critical than in any other culture. Feedback provides external verification and tells the competence culture whether its output has merit.

Life in a competence culture is intense and high-strung. People always have standards to reach and go beyond. The work is rigorous and carries a sense of urgency. The norms are excellence, superiority, and challenge.

The competence culture has its own brand of epistemology. People believe something when conceptual system goals and reality connect. Conceptual systematism holds sway—theories, concepts, and principles. As a result, the framework for information and knowledge is built essentially around the conceptual system goals of the organization and the extent to which those goals are met.

The competence culture has competition at its center. People compete with one another in a highly competitive world which has more stringent standards and problems that stand in the way of getting the job done. The more competent you are, the more effective you are at competing. In contrast to the collaboration culture, this culture is often a win-lose world in which discord is present and less competent people fail. Because superiority is paramount, inferiority stands out and is discredited. An organization whose goal is to create one-of-a-kind products or services instinctively fills the organization with one-of-a-kind people—specialists—to achieve that goal.

Competition entails striving to go beyond set standards. It is inherently future oriented and possibility oriented. The process for competition must be objective, prescribed, and external; it cannot be subjective, evolutionary, or open-ended because people would have no way of knowing how to determine who won and who lost.

Digital Equipment Corporation (DEC) in Maynard, Massachusetts, typifies a competence culture. *The New York Times* profiled DEC and its founder and CEO, Kenneth Olsen, in 1992.

> Last year, John Rose, a 16-year Digital veteran who spent several years leading the company's catch-up efforts in the personal computer market, was stunned to discover that Mr. Olsen had created a competing group that was pursuing a vastly different strategy for the company . . . that a personal computer designed and built in Taiwan would become Digital's predominant personal computing product. Mr. Rose tried to start discussions about what the right approach for the company might be, but he was rebuffed by Mr. Olsen . . . The episode was quintessential Digital under Mr. Olsen, who has a penchant for squaring off competing activities to spur innovation. "If you don't have competition pressuring you, you just won't move," Mr. Olsen said. Clearly, this is an example of a culture that prizes competition that is external to people (e.g., Mr. Rose, among others) and that is possibility oriented.
>
> "In the 1980's Digital had a clear message," said Stephen Smith, a computer industry analyst at Paine Webber . . . "These days they are

throwing a lot of balls in the air and seeing if they can catch one. They've gone from one architecture to multiple architectures, from a single operating system to multiple operating systems. They seem rudderless right now, and customers are very confused."[8]

Digital is not a collaboration culture teaming up with customers. It may be rudderless, as Smith suggests, or it is trying to build the best system it can in order to have the most competent and best niche in the marketplace. Olsen has said that he would run DEC only as long as it was the best.

Because of Olsen's background as a research engineer at the Massachusetts Institute of Technology, the probability is high that he operated DEC from the paradigm of a university, and that the individual motive of achievement served as his fundamental model for an organizational culture within the company.[9]

DEC's way to success has always been characterized by innovation, possibility, and striving to be the best in the business. Unfortunately, its "problems appear to be as complex and confusing as its legendary matrix organization, in which layers of management disseminate mountains of data seeking consensus within Digital."[10]

The Cultivation Culture

The cultivation culture is a possibility–personal culture. It pays attention chiefly to potentiality, ideals and beliefs, aspirations and inspirations, and creative options. Its decision-making method is people driven, organic, open-minded, and subjective.

If you could transform a cultivation culture into a working person, it would be best represented by a religious steward, such as a minister, priest, or rabbi. Religious stewards focus on catalyzing and cultivating growth and development among their people. They strive to help people fulfill their potential, particularly spiritual potential. They focus on people's inspirations and aspirations, and herald ideals and higher level purposes. The content of the steward is possibilities and potentialities. The process used is highly subjective and organic, and frequently emotional. A religious steward emphasizes empowering people and unleashing that which is internal to them. This occurs both in the steward's own organization and with outside constituents.

The cultivation culture is very much a world of purpose, evolution, and change. Its organizations are magnetic—the magnetism caused essentially by the level of commitment that its people can attain. The cultivation culture gets results because its people believe in the purposes of the organization, in themselves and their leaders, and in enriching their customers or constituents. Willpower makes things happen in this world.

The culture is value centered. Values and the value of people hold sway. Self-expression is highly encouraged, indeed nourished. People are given every opportunity to be all that they can be, to be possibilizers. They identify strongly with their organization.

The epistemology in the cultivation culture is determined by the values that are at its center. People know or believe something when there is a connection between what they value and reality; when what is espoused is put into operation. Knowledge making in the collaboration culture is based on experience but in the cultivation culture it is based on what people value or hold as ideal.

Change is perhaps more automatic and natural in this culture than in any of the other three. Growing, creating, and developing are essentially matters of change.

In clear contrast to the objectivity of control cultures, the cultivation culture is one of subjectivity. Decision making hinges on content and processes that exist within people. It is a highly internal world. Organic, evolutionary, and emotional judgment making rest within people. As a result, people within a cultivation culture work in the service of internals—what is worked on and how it is done.

Herman Miller, Inc., the furniture maker, is a superb example of a cultivation culture. Max DePree, chairman and CEO of Herman Miller, writes about leadership, organizations, and people.

Leaders owe a covenant to the corporation or institution, which is, after all, a group of people. Leaders owe the organization a new reference point for what caring, purposeful, committed people can be in the institutional setting . . . what we can do is merely a consequence of what we can be. Corporations, like the people who compose them, are always in a state of becoming. Covenants bind people together and enable them to meet their corporate needs by meeting the needs of one another . . . a corporation or business or institution [is] a place of fulfilled potential . . . [whose] effectiveness comes about through enabling others to reach their . . . personal potential and their corporate or institutional potential.[11]

DePree adds that issues of the heart and spirit are important to the families, the workplace, and the extracurricular activities of the people at Herman Miller. "We are emotional creatures, trying through the vehicles of product and knowledge and information and relationships to have an effect for good . . . Deep in who we are today lies a challenge waiting. It is not an external mystery— the question of what we can be lies within us, for whatever we do expresses the character of the people who are this company."[12]

CULTURAL OPPOSITES

Cultures on the diagonal to one another in Exhibit 7–1 are opposites. Actuality–impersonal contrasts most sharply with possibility–personal; actuality–personal contrasts with possibility–impersonal. The control culture differs most from the cultivation culture and the collaboration culture differs most from the competence culture.

When two cultures differ from one another in both content and process, they are opposites. At a minimum, this suggests that leaders who plan to change their culture would make more headway by moving toward one of the two adjacent cultures with which there is either a content or process alliance. For example, a control culture will find it easier to try to become either a collaboration or a competence culture.

Capital Holding's Direct Response Group (DRG) illustrates this principle. DRG, a longtime control culture and successful direct marketing insurance company, determined in 1987 that its business and market was changing drastically. Its president and CEO, Norm Phelps, decided to move the company from a control culture to a collaboration culture. He and his organization have been engaged in a massive cultural change ever since. Interestingly, the *one* area that has proved the most difficult to change is company decision making, particularly sharing the power of decision making. Donald Kennedy, senior vice president in charge of implementing team-based work process change, says "sharing decision-making power is the chief obstacle. In the past, executives were rewarded for giving directions and making decisions. The team concept implies something very different: that no one person has the best answer, but that the best answer is developed from a variety of viewpoints."[13] DRG's decision to move from a control to a collaboration culture is fundamentally a move along the process axis from impersonal decision making to personal decision making. This move is a change in how an organization makes decisions. Had Phelps and Kennedy been aware of our model, they could have anticipated and systematically implemented initiatives for change in decision making aimed at DRG's needs.

THE ESSENCE OF CULTURE FORMATION

What emerges from this material is that a culture is essentially formed from (1) what it pays attention to and (2) how it makes judgments and decisions. In addition, it is important to keep in mind the two kinds of content and the two kinds of process. It is not just any kind of content or process, but actuality versus possibility content and impersonal versus personal process.

Founders and leaders will benefit greatly by keeping these distinctions in mind when they start building an organization. In addition, they will find it helpful to determine whether they want to form an actuality or possibility organization for their organization content and an impersonal or personal organization for their organization process. The choice they make early on in will have much to say about the kind of core culture they establish within their organization.

As their organization grows and develops, the paradigm (i.e., content and process) that they choose will serve as a powerful guideline for a coherent and integrated maturation of their organization. It will allow them to keep their culture focused and consistent. The paradigm serves as a road map to help them stay on course and prevent them from veering off course onto subcultural back roads that will only cause confusion.

Leaders of existing organizations will find this information helpful in determining the kind of core culture that already exists. If their culture lacks coherence and integration, this information should help them discover where they are offtrack, what they need to keep building, and what they need to change or discontinue.

Once leaders determine the core culture they have in their organization they can benefit by focusing separately on their areas of content and process. Where the content is incomplete or inconsistent, changes can be made that will foster their organization's effectiveness. They can conduct the same effort with process by taking a closer look at how their organization makes judgments and decisions and, then, make any necessary changes.

NOTES

1. Shils takes the position that societies have cultural centers and peripheries. A set of ultimate ideologies appears to emerge consistently at the center where there is considerable consensus about each one although they may conflict with one another. These in turn radiate outward from the center toward the diverse segments of the periphery in varying degrees of consensus. See E. Shils, *Center and Periphery: Essays in Macrosociology* (Chicago: University of Chicago Press, 1975); and E. Shils, "Center and Periphery: An Idea and Its Career, 1935–1987," in L. Greenfield and M. Martin, eds., *Center and Periphery: An Idea and Its Career* (Chicago: University of Chicago Press, 1988).

2. These two axes (actuality–possibility and personal–impersonal) coincide almost exactly with the functions or modes of orientation described by Carl Jung in *Psychological Types* (London: Routledge and Kegan Paul, 1923). The actuality–possibility axis mirrors Jung's "sensation–intuition" functions and the personal–impersonal axis mirrors his "feeling–thinking" functions. Jung's work on psychological types has been carried on, particularly by I. B. Myers, *Introduction to Type*, rev. ed. (Palo Alto, CA: Consulting Psychologists Press, 1987), and I. B. and P. B. Myers, *Gifts Differing* (Palo Alto, CA: Consulting Psychologists Press, 1980). S. Hirsh and J. Kummerow's *Lifetypes* (New York: Warner Books, 1989) is currently enjoying considerable popularity and utility in individual and organizational development. Another close parallel can be found in the work of two sociologists, F. R. Kluckhohn and F. L. Strodtbeck, *Variations in Value Orientations* (New York: Harper & Row, 1961). These two authors conducted a

massive comparative study of a number of cultures in the US Southwest. They noted that cultures make different assumptions about how to act, which reflect their assumptions about human nature and the fundamental relationship of the group to the environment. At one extreme is a "doing orientation," which correlates closely with (1) the assumption that nature can be controlled and manipulated, (2) a pragmatic orientation toward the nature of reality, and (3) a belief in human perfectibility. The doing orientation focuses on the task, efficiency, and discovery. Kluckhohn and Strodtbeck also describe a "being-in-becoming orientation" that emphasizes self-development, self-actualization, and fulfilling one's potential.

3. C. F. Knight, "Emerson Electric: Consistent Profits, Consistently," *Harvard Business Review* (January–February 1992), pp. 58, 62–63.

4. Ibid., p. 60.

5. R. Levering and M. Moskowitz, *The 100 Best Companies To Work For in America* (New York: Currency/Doubleday, 1993), p. 60.

6. Ibid., p. 62.

7. Ibid., p. 63.

8. G. Rifkin, "The Hand That Shakes Digital Up," *New York Times* (May 15, 1992), sec. C, p. 1.

9. In the fall of 1992, the board of directors at DEC asked Olsen to step down as chairman and discontinue day-to-day management. Whether Digital will remain a competence culture or shift away from the leadership approach so thoroughly inculcated by Olsen remains to be seen.

10. Rifkin, sec. C., p. 5.

11. M. DePree, *Leadership Is An Art* (New York: Dell, 1989), pp. 15, 19–20, 68.

12. Ibid., pp. 87–88.

13. R. S. Buday, "Forging A New Culture At Capital Holding's Direct Response Group" *Insights Quarterly* 4, no. 2 (Fall, 1992), p. 48.

Chapter Eight

A Framework for Developing Your Organization

The concepts of the four core cultures lead directly to a set of implications for organizational development, leadership and management development, individual and organizational change, career and performance management, and numerous other areas. To pursue even one of these areas in depth goes well beyond the scope of this book.[1]

This chapter provides the reader with a beginning framework for organizational development that draws on the concepts learned about the four core cultures. Organizational development is the logical place to start. In doing so, I will illustrate, in skeletal fashion, how an understanding of the four core cultures can substantively affect your approach to developing your own organization.

If you peel back the layers of any organization, you will find one of the four core cultures. An organization may have some of the manifestations of other core cultures, but at its center it has just one. More than one core culture in an organization would be too chaotic and confusing for people to tolerate. In the very early beginnings of an organization's life, it may have more than one core culture. But after any meaningful period of time, that organization organizes itself and forms only one core culture. A conglomerate composed of several separate organizations may have several core cultures, but each organization *within* the conglomerate has its own separate core culture. Some of these separate organizations could have the same core culture, but each is nonetheless a singular core culture.

For our purposes, the focus is the majority of organizations that are reasonably healthy and successful, and which have been functioning for some period of time.

ANALOGUE OF CULTURE AND CHARACTER

Organizations are *organisms*—the word *organization* is derived from the word *organism*—that have essential qualities about them paralleling those of individual human beings. After all, organizations are built and operated by human beings, so it is inevitable that they correspond to their organismic counterparts. You have read in the preceding chapters that each core culture has a unique socialization base and unique individual motivation base. These two foundations are human in nature. Like a human organism, an organization attends to certain matters and not to others; it forms judgments and makes decisions; it has a unique style; it resists or embraces change; and it strives to succeed.

An organization's core culture corresponds closely to an individual's core character. Both organisms naturally develop a more or less integrated persona or constitution that provides a paradigm for working, living, and striving to succeed. As you recall from Chapter 1, a paradigm is a framework for how things ought to be. Both organizations and individuals rely greatly on paradigms. Because they have this similar need, culture and character are analogues of one another. Culture is important to an organization and character is important to an individual because both (1) provide consistency, order, and structure; (2) establish an internal way of life; and (3) determine the conditions for internal effectiveness. Organizational culture is "the way we do things in order to succeed" and individual character is "the way I do things in order to succeed."[2]

To organize means to provide with an organic structure; to arrange in an orderly way; to make into a whole with unified and coherent relationships. Both organizations and individuals organize themselves, organizations by developing a culture and individuals by developing a character.

The parallel between organizations and individuals and organizational culture and individual character is important for a number of reasons. First, it partly underpins the rationale for an organization having only one core culture; for the same reason, an individual has only one core character. For both, the whole process of being organized is to build a structure. Culture and character serve this purpose. Individuals and organizations that have more than one core character are disorganized, unstructured, incoherent, and disunited—they are lost without a road map or paradigm.

You learned in Chapter 7 that a culture is formed essentially from what it pays attention to and how it makes judgments and decisions. Therefore, if an organization organizes by way of its culture, it also organizes by the same content and the same process.[3]

The parallel between organizations and individuals and between organizational culture and individual character is important for a second reason, which is the essential message of this chapter. As organisms, an organization and an individual person parallel each other in terms of true effectiveness; what effectiveness means for one is analogous to what it means for the other.

ORGANIZATIONAL EFFECTIVENESS

Individual effectiveness is essentially equivalent to balanced integrity. The same holds true for an organization.[4]

Integrity means to be complete, whole, or unified. Balanced means to be in a state of equilibrium. For our purposes, together they mean (1) that the individual or organization has reached a level of completeness and that no further development is possible, and (2) that the organism is stable and capable of keeping internally diverse elements in balance. Realistically, however, the full accomplishment of this state of effectiveness is highly unlikely, if not impossible.

Fortunately, we are all in the same boat. The definition of effectiveness reaches the real nature of life because it is rare that we arrive at a point where we are fully integrated and. balanced. Striving to be more effective is more a journey than a destination. Nevertheless, the closer an organization gets to balanced integrity, the more effective it is.

Interestingly, the definition of organize is closely related to the notion of integrity: to make into a whole with unified and coherent relationships. The unified, integrated organization is more effective. And, as we have seen, organizing is accomplished by the very process of forming a culture.

The process of culture formation is natural and automatic; an organization forms a core culture and an individual forms a core character. A central premise of this book is that humankind has derived four such core cultures.

There is no single core culture that every organization should emulate. J. O'Toole's argument that the creation of a single corporate culture for all companies has been wrongheaded[5] coincides with my thesis that organizations, in the United States particularly, have been done a disservice by those who insist that they must model themselves after what the Japanese or "excellent" companies are doing.[6] As you have read, striving for excellence in itself is the province of the competence culture and centering organizational activity on teams is inherently the province of the collaboration culture. It may well be that one kind of core culture is naturally suited to a certain kind of endeavor, but that is certainly a different premise than one that insists that every organization should have one kind of core culture.

The key issues concerning organizational effectiveness are the following:

- How clear are people about the true nature and strengths of their core culture?
- How integrated is that core culture?
- How much wholeness or unity has been reached?
- How balanced is the core culture?

These four questions comprise the fundamental elements of a strategy and process for developing one's organization.

MEASURING AND DEVELOPING YOUR ORGANIZATION'S EFFECTIVENESS

Use the four questions as your fundamental strategy and process for developing your organization's effectiveness. The following steps provide a systematic framework for implementing such a program.

Step 1: Determine Your Core Culture

By now you should have completed the Core Culture Questionnaire (see Chapter 2) and determined your organization's core culture. Now it is a good idea to add an important step in your use of the questionnaire: Enlist as many people in your organization as possible to fill out the questionnaire independently. If this is not possible, ask a well-stratified sample to do so. The value of doing this is that it helps in building a consensus about which core culture exists within your organization. This will pay great dividends in implementing your program of organizational effectiveness. Another value of having a large number of people complete the questionnaire is that it adds weight to your overall results.

If the collective results do not yield a consensus on your organization's core culture, establish discussion groups in which people can discuss in-depth where they agree and disagree. The discussion groups should be comprised of a stratified sample of people who represent the organization's diversity and who have been with the organization for a substantial time. Probing for examples of actual organizational behavior should eventually help everyone reach agreement.

Remember that the goal is self-awareness, not arriving at bad and good judgments. No one culture is better than another.

Step 2: Capture Your Culture's Strengths

Once your organization's core culture has been determined, you can delineate its strengths. These will serve as the essential building blocks of any development or change that you embark upon.

Begin with Appendix A in the chapter on the core culture that represents your organization (see chapters 3–6). The appendix depicts the strengths of your organization's core culture. You also will find it helpful to review the tables in Chapter 7. They spell out the inherent natural strengths of each core culture from a strategic, leadership and management, structural, power, and relationship focus. Rereading the chapter on your core culture may also provide new insights.

The next step is to assemble your top managers and create a collective statement of your organization's cultural strengths.

This statement should reflect the true reality and collective experience of your organization. It is important that all the top management in your organization can acknowledge and support the collective statement.

You will find that the statement of cultural strengths will serve as a guide for moving your organization into the future. It will help your organization's top managers stay proactively focused so they can make the right decisions about fitting new elements into your culture in the future. Any new developments or changes that your organization attempts will bear much greater fruit if they are designed and implemented with these strengths in mind.

The essential value of the statement is that it will keep your organization's strategic and operational priorities at the forefront of management thinking. The more that it is used as an ongoing frame of reference, the better management can keep the organization on course and shorten the time frame for accomplishing what needs to be done.

Step 3: Determine Your Core Culture's Level of Integration

Level of integration means the following:

- The degree of coherence between the elements of your organization.
- The extent that the elements of your organization are aligned with one another.
- The degree of focus within your organization.

Organizational elements can be a department, a process, an individual, a committee, a way of making decisions, a key norm; a leadership approach, a management principle, a structure, a project—in short, any part of the life of your organization.

The first thing to do in Step 3 is to gather about eight to ten people who know your organization well and who typify its structure. Ask this group to analyze systematically the organizational elements depicted in Exhibit 8–1. The three principal components of balanced integrity—integration, wholeness, and

EXHIBIT 8–1
Balanced Integrity Lens System

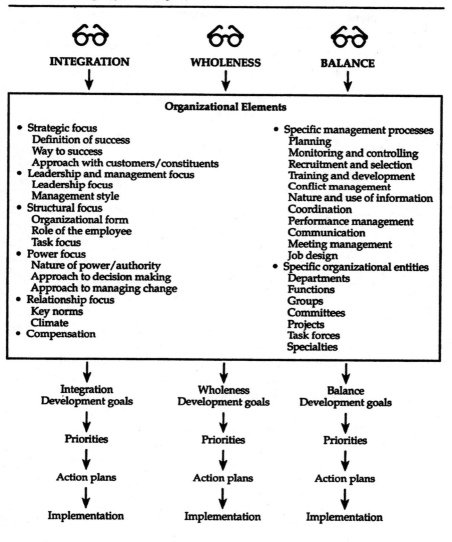

INTEGRATION   WHOLENESS   BALANCE

Organizational Elements

- Strategic focus
  Definition of success
  Way to success
  Approach with customers/constituents
- Leadership and management focus
  Leadership focus
  Management style
- Structural focus
  Organizational form
  Role of the employee
  Task focus
- Power focus
  Nature of power/authority
  Approach to decision making
  Approach to managing change
- Relationship focus
  Key norms
  Climate
- Compensation

- Specific management processes
  Planning
  Monitoring and controlling
  Recruitment and selection
  Training and development
  Conflict management
  Nature and use of information
  Coordination
  Performance management
  Communication
  Meeting management
  Job design
- Specific organizational entities
  Departments
  Functions
  Groups
  Committees
  Projects
  Task forces
  Specialties

Integration Development goals	Wholeness Development goals	Balance Development goals
Priorities	Priorities	Priorities
Action plans	Action plans	Action plans
Implementation	Implementation	Implementation

balance—can serve as lenses for the group to sharpen analysis and magnify the details of a problem. The lenses will help you see more clearly which of the organizational elements needs to be worked on to develop your organization's effectiveness.

The questions in Exhibit 8–2 help you deal specifically with your core culture's level of integration. Your group of analysts should answer each question as it relates to each of your organization's elements (see Exhibit 8–1). If your organization has elements that bear attention but which are not depicted in Exhibit 8–1, be sure to include them in the analysis.

The answers determined by the group will provide you with level of integration development goals that you can then set out to accomplish. You should determine the priority of your development goals and establish action plans for each before implementing them. The actual plans and processes for accomplishing each goal will vary, but essentially the process will be one of (1) changing something that is present or (2) taking out something that doesn't belong.

You should find it helpful to review the list of strengths generated in Step 1, the tables in Chapter 7, and appendixes A and B at the end of the chapter that describes your core culture. In addition, it is helpful to review the content and process of your core culture and to consider how congruent the behavior of your organization's is with each.

Finally, any organization with one core culture can have operating within it one or more of the characteristics of another core culture. Most organizations operate this way. However, this is essentially an integration issue, and there should be no problem as long as the other core culture characteristics function in the service of your core culture. Problems arise when an organization behaves as though it were possible to have more than one core culture or adopts other core culture characteristics for nonintegration purposes. SOS, Inc., (see Introduction) ran into difficulty because its president forcefully tried to integrate two separate core cultures within the same organization.

Step 4: Determine Your Core Culture's Degree of Wholeness

Wholeness has to do with completeness. As you recall, it is not a good idea to hold out completeness as a desirable end point for your organization. In and of itself, it is impossible to accomplish.

EXHIBIT 8–2
Balanced Integrity Lens System (Level of Integration Lens)

- Where are the elements of our organization inconsistent, given the nature of our core culture?
- Where do we behave in a manner that is incongruent with our core culture?
- Where do we give ourselves mixed messages? We are this type of core culture, but what are we demanding from people that is incongruent with our core culture?
- Where do we belie the nature of our core culture?
- Where are we at cross-purposes with one another? Where are we uncoordinated?
- Are we aligned with our natural and automatic definition of success?
- Do we have a subsidiary organization that has a different core culture from ours? Are we trying to make it into our own image?
- Where are we trying to be all things to all people? Where are we buying into fads that are really elements of another core culture? Are we trying to institute qualities, approaches, or processes that are incongruent with our core culture, but which everyone is touting as the best thing to do?
- Where do we lack internal harmony?
- Where are we trying to have our organizational cake and eat it too?
- Where do we keep going off on tangents?
- Where do we lose concentration?
- What is present within our organization that does not fit? What do we have that doesn't belong? What keeps getting in the way of our core culture?
- Given our core culture, is our way with customers and constituents the same or parallel to our way internally?

Nonetheless, wholeness is a fundamental component of your organization's effectiveness; it is definitely something to work hard at approximating.

Ask the same group of analysts to follow the same process used in Step 3, but this time on the level of wholeness lens (see Exhibit 8–3). They should ask questions about the level of wholeness for your overall organization and each of the organizational

EXHIBIT 8–3
Balanced Integrity Lens System (*Level of Wholeness Lens*)

- Given our core culture, where are we incomplete?
- What is missing that belongs in our culture?
- What don't we have operating within our core culture that we need?
- What more do we need to do?
- Where do we lack unity? What has been broken apart that needs to be put back together?
- What is damaged or in disrepair? What is disassembled that should be reassembled?
- What are we omitting outright or deciding not to do?
- What knowledge, skills, attitudes, or motivations do we need?
- Given our core culture, what is unfinished?

elements (see Exhibit 8–1). Again, other organizational elements not shown may need to be addressed by the group.

The answers will provide you with level of wholeness development goals that you can begin to work on. As in Step 3, you will need to set priorities for goals and establish plans of action for each before implementing them. The goals determined by the group will address what needs to be added to your organization and its operations. You will have to decide whether your organization has the resources to make such additions. If so, you can formulate and implement plans of action. If not, you will have to defer implementation until you can acquire them.

The strengths of your organization generated in Step 1, the tables in Chapter 7, and the material on your core culture's content and process will again be helpful in working through this step. Your group of analysts will find it helpful to reread the chapter on your core culture with its description of what a complete core culture such as yours looks like.

Sigma Electric began having problems when a new CEO's leadership caused the organization to be incomplete. Because it was a control culture, Sigma needed an authoritative approach to leadership at the top.

EXHIBIT 8–4
Balanced Integrity Lens System (Degree of Balance Lens)

- Given our core culture, where are we out of balance?
- Where are we operating in excess? Where are we overdoing things? Where have we allowed things to go too far?
- What are we unnecessarily overemphasizing or underemphasizing?
- Where are we exaggerating our core culture's natural strengths?
- Given our core culture, where are we getting too greedy?
- Where are we taking shortcuts?
- What specific elements from other core cultures do we need to incorporate into our organization to be in balance and more effective?
- Where are we refusing to allow natural and necessary tensions in our organization to continue? Where are we imposing one way of doing something when we should be managing in more than one way?
- Where do we lack stability?
- Where are we out of proportion?
- Where do we lack reciprocity?
- Where have we lost our adaptability?
- Where do we need more equality?

Step 5: Determine Your Core Culture's Degree of Balance

The final step in your pursuit of increased organizational effectiveness is to work on the degree of balance within your organization. Balance means the extent to which your organization is operating in a state of equilibrium.

The process recommended is identical to that of Steps 3 and 4. Ask your group to analyze your organization and its elements (see Exhibit 8–1) utilizing the degree of balance lens (see Exhibit 8–4). Remind them to include applicable organizational elements not listed. Encourage the group to review the strengths of your organization generated in Step 1, the tables in Chapter 7, and the appendixes at the end of the chapter on your core culture. Their answers to the questions in Exhibit 8–4 will provide you with your organization's degree of balance development goals.

It is reasonable to predict that you will incorporate elements from one or more of the other core cultures into your organization's culture. A close look at cultural opposites shows why. The two sets of opposites are control and cultivation on the one hand and collaboration and competence on the other. Opposite cultures, by and large, have juxtaposed strengths and weaknesses, that is, the strengths of one are often the weaknesses of the other. You might derive considerable benefit from looking at your opposite culture to see where you are possibly underemphasizing something important for your organization's effectiveness.

For example, control cultures tend to underemphasize empowering their people—a hallmark characteristic of cultivation cultures. After examining its opposite culture, a particular control culture might implement an ongoing process for eliciting more employee input about how things should get done. Or a competence culture could implement processes for teaming among managers or employees that enhance the effectiveness of the organization.

In all of this, it is important to be as specific as possible: Determine exactly what is needed, where and how it needs to be implemented, and who should be in charge of the operation. Sweeping generalizations that your organization needs to be more collaborative or more competitive won't work. It is also important that any elements borrowed from other core cultures be incorporated into your organization in such a way that they function in the service of your core culture.

Change Technologies, the management consulting organization (see Introduction), went out of existence because it got so out of balance. The central strengths of its cultivation culture nature were allowed to run to excess. Given the talent within Change Technologies, the company might be thriving today if its people had kept the company more balanced by adopting some elements from the opposite core culture, the control culture.

THE PROGRAM IN ACTION

My colleagues and I implemented the organizational effectiveness program with Consumer Plastics, Inc. (a pseudonym), a consumer products manufacturing company with annual sales of $80 million.

In Step 1, we administered an early version of the Core Culture Questionnaire to all company personnel; 85% of those responding identified the company as a control culture. Collaboration culture received the second highest score, particularly in the organization of work and promotional practices.

Before implementing Step 2, we put together a stratified sample of 25 people who represented a microcosm of the whole company. Starting from its control culture nature, this group generated a list of Consumer Plastics' cultural strengths. Their delineation was later revised and subsequently approved by senior management. We often turned back to this list as we worked our way through the other three steps.

In Step 3, the group addressed Consumer Plastics' level of integration. Each person in the group answered each question individually. The group then arrived at a consensus that the company lacked integration in two clear areas.

The company's approach toward Outdoor Equipment Company (another pseudonym), a small outdoor camping products company that Consumer Plastics had acquired six months earlier, was unintegrated. It was clear to the group that Outdoor Equipment Company was a competence culture and that Consumer Plastics had been vigorously trying to impose its control culture ways on Outdoor Equipment. It was equally clear that this managerial approach was not working. The group's solution was fairly straightforward. It proposed that Consumer Plastics back off and allow Outdoor Equipment to function as a competence culture. This recommendation was subsequently endorsed by Consumer Plastics' senior management and eventually implemented.

A second area in which the company lacked integration was identified in part from the Core Culture Questionnaire. Consumer Plastics was a collaboration culture in two areas—work organization and promotional practices. Company practice was to ask people to function as generalists and only those people who exhibited generalist capabilities were eligible for promotion. The group determined that people in the company were being given a double message about what was expected of them and what it took to advance within the company. As a control culture, Consumer Plastics' structure was hierarchical and company functions were clearly delineated. Managers of those functions were understandably anxious to retain their performers, but they also felt compelled to move people across functions with considerable frequency because of company policy. The upshot of all this was that no one's heart was in the policy because it was at cross-purposes with the culture. The group recommended that the policy be changed to reflect the natural functionalist emphasis of the company's control culture. Senior management subsequently agreed to this recommendation.

The group next went to Step 4 by asking and answering questions on the company's degree of wholeness. Again people addressed these questions individually and later the group arrived at a consensus. It was clear that work was needed on one area—a delicate one dealing with the president's leadership style. The president had served in his position for one year; [he was only the third president in the company's history.] While everyone regarded him as congenial, bright, committed to the company and its people, and a capable businessman, he also was regarded as too much of a "relator" (the group's term) and not enough of a "director." The group certainly wanted the president to retain his concern for people, but they also stressed the need for him to take command of the organization and to function more authoritatively. The group's decision was to give the president time to think through this issue for himself and the company. The president agreed to think about it. After considerable soul-searching and private discussions with others, he eventually agreed with the findings of the group and embarked on a personal development program designed to change his leadership style.

In Step 5, the group addressed the company's degree of balance. The consensus of the group was that the company had to work on nine developmental goals. It is still addressing a few of them. Let's concentrate on the three goals that the group decided had the highest priority.

First, the group determined that the company had grown too rigid in its design of work and in the deployment and promotion of its people. Most people in the company were stagnating and becoming rather hidebound. The generalist policy referred to in Step 3 was an attempt to address this issue. The group designed a program for developing people that was in keeping with the company's control culture nature yet also made great strides in fostering development. This program, subsequently endorsed by senior management, called for the following actions:

- Vigorous job rotation assignments *within* functions.
- A systematic rotation of job assignments *across* functions—but only after essential functions were properly staffed and after fully capable backups were trained and ready to go.
- Design of a functionally oriented benchmarking program whereby people from each function in the company visit other companies that have a reputation for highly effective functions of one kind or another. The information is brought back to Consumer Plastics for consideration and possible incorporation within the appropriate function.

Second, the group determined that more collaboration and teaming at the top of the organization was needed, particularly in the making of business strategy. The group believed that the environment of the company was changing at a faster rate than it had in the past. Various functions—notably manufacturing, marketing, and sales—were experiencing a more volatile marketplace, more intense competition, and a much wider array of competitive products. The group's determination was that the functionalism of Consumer Plastics was too extreme and, as a result, the company was not adapting effectively to the marketplace. The real problem, the group decided, was the company's difficulty in getting the best *collective* thinking from all functions within the company, particularly manufacturing, marketing, and sales. Accordingly, the group recommended that a cross-functional business strategy "council" be formed and implemented, and senior management sanctioned it. Interestingly, the process itself—the organizational effectiveness program—was instrumental in fostering agreement by top management. It is an example of how beneficial getting people together to address company-related issues can be.

The third priority concerned too much upward delegation of responsibility. The group estimated that people at lower levels were overcomplying with their supervisors and managers and did not take enough personal responsibility for their own work. Almost anything out of the ordinary experienced by employees at lower levels was much too quickly and willingly delegated upward for supervisors and managers to handle. The group's recommendation was to set up a training program that would address the issue of upward delegation and train the employees at lower levels how to take more responsibility for their work. Senior management did not agree with this recommendation and returned it to the group for reconsideration. Senior management's perspective was that, while this was a developmental goal for the company, focusing only on employees and not on supervisors and managers would not accomplish what the company needed. When the recommendation was later revised to include supervisors and managers, senior management eventually sanctioned it.

IMPLICATIONS FOR
ORGANIZATIONAL CHANGE

The nature of organizational core cultures, culture formation, and organizational effectiveness suggests a number of implications worth considering.

First, as Shakespeare said, it is critical to know thyself. In *The Road Less Traveled*, M. Scott Peck reasons that for individuals to grow, develop, and change, they must engage in a "continuous and never-ending process of stringent self-examination."[7] The same holds true for an organization. The more fully an organization knows itself, the greater its potential for positive change and increased effectiveness.

Second, change has to do with being true to oneself. In organizational terms, change relates to being true to your core culture. Change comes from within. Effective change builds on the existing culture; an organization turns inward and builds on its strengths. Carl Jung said of change and development: "If a plant is to unfold its specific nature to the full, it must be able to grow in the soil in which it is planted."[8] An organization does not change by trying to imitate other organizations. Change, continued development, increased effectiveness, and added success all come from attending to and building on your organization's internal paradigm.

A. L. Wilkins and N. J. Bristow, who have worked on developing organizations for many years, take a similar perspective. They stress the importance of organizations "honoring their pasts" and developing organizational culture "without destroying it."[9] John Kotter and James Heskett's research points in the same direction. Between 1987 and 1991, they investigated 207 organizations to determine whether organizational culture was critical to economic performance and, if so, why. They concluded that culture was indeed critical to economic performance and that successful organizations were not only strongly committed to core values and behaviors but also they talked endlessly about them and did not allow new managers to undermine them.[10] Peter Drucker, an authority on organizations and their management, says that "culture . . . is singularly persistent . . . In fact, changing [organizational] behavior works only if it can be based on the existing 'culture.' "[11]

Third, change takes time. It rarely happens overnight. And it is often painful and disrupting. When you consider the true nature of organizational effectiveness, we never really arrive. An organization is never cooked. Pushing for more and more effectiveness is a continuous process. There is no clear end point marking that we have finished.

Fourth, cultural change is possible but very difficult. Trying to go from one core culture to another is quite an undertaking, analogous in some ways to changing an individual's core character.

> People frequently change their opinions, their behavior, sometimes even the values and beliefs they espouse, but their personalities remain essentially unchanged. The same is true of companies. Psychologists have always been leery of sudden conversions, because they usually prove quite superficial and temporary. Permanent changes are most often very gradual and rarely radical . . . Deep-rooted changes [in organizations] are not brought about merely by a change in the goals top management seeks, the strategies it adopts, or the actions it initiates.[12]

After looking at his own and others' research for the last 25 years, Denison concludes the following when it comes to cultural change: (1) Culture change is pushed "in response to the demands of the business environment . . . (it) was most often driven by a crisis of mission and strategy and the need to adapt, rather than by any intention to change the internal organization itself"; (2) Culture change typically means "new players, not the conversion of old players"; and (3) Cultures have "tremendous inertia and change very slowly. Culture is by definition a collectively internalized normative system that outlives any one individual . . . A decade is a short period of time in which to expect to institutionalize cultural change within a large organization."[13]

The leaders and managers of an organization will make much more headway preserving their core culture and working vigorously for more integration, completeness, and balance than by trying to make it over into a completely different culture. If, nonetheless, they persist in trying to change their culture, they should move toward one of the two core cultures with which their culture has either a content or process alliance.

The deepest issue concerning the need for organizational and cultural change is the extent to which a particular organization is congruent with the true nature of its enterprise. Leaders drive culture formation and maintenance; how they perceive their environment and the strategy to influence that environment is a key ingredient to the kind of core culture that is established. If their perceptions are incorrect, then they establish a culture that is fundamentally incongruent with their organization's true role in the environment. When this happens, ineffectiveness is built in at the start and the organization labors under a rubric for success that is structurally flawed. Then it would appear that fundamental cultural change is required for long-term survival and success.

The public school system in the United States appears to be an example of this phenomenon. It was originally modeled on the factory; children were seen as products to be processed in an assembly-line kind of framework, which lent itself to the development of a control culture paradigm. However, the true nature of the elementary and secondary education system points more naturally toward the formation of a cultivation culture at its core. Those crying for change in the public school system in the 1990s should look to the nature of the cultivation culture for direction.

IBM's travails are another illustration of this principle. Mark Stahlman argues that IBM's one-size-fits-all business model no longer works in the computer industry.

During the past 30 years, computer-based technology has fueled the establishment of four major computer industries. These industries have been "pulsed" out of the technology like bursts of electromagnetic energy from a rapidly spinning neutron star. But while each pulse comes out of the same fast-evolving technology base, each is radically different in its basic economic structure. Every aspect of these computer industries—from who supplies the basic technology to how products are configured, sold and serviced—has been completely reinvented from one pulse to the next. The inability of IBM's managers to recognize this fact and create new business models appropriate to the distinct markets of the many computer industries explains IBM's failure.[14]

The four major computer industries—the mainframe computer industry, the minicomputer industry, the personal computer industry, and the workstation computer industry—correspond roughly to the four core cultures.[15] Recasting Stahlman's position in the framework of corporate cultures, IBM ran into difficulty because its leadership tried to import IBM's mainframe control culture into three other, inherently incongruent, computer industries—minicomputers, personal computers, and workstation computers. Each of these requires a separate, integrated, balanced, and whole core culture.

In any industry, the better-performing organizations had cultures that were congruent with their environment. Wholesale culture change is required when an organization's core culture is inherently inconsistent with the true nature of its enterprise.

NOTES

1. Additional works designed to expand more fully on these implications are planned by this writer. Currently in development is a comprehensive and systematic program for organizational and management development which utilizes, among other materials, results from a newly developed and validated organizational culture survey

that yields a picture of what an organization's core culture is and what sub-factors in the organization are congruent (or incongruent) with that core culture.

2. Frederick Harmon and Garry Jacobs place considerable emphasis on the parallel between individual character and organizational culture: "Individual and corporate personality are constituted in much the same way. Both are living forces characterized by energy and direction . . . Each corporation has a psychic center, too, which consists of the beliefs, values, mission, attitudes, and objectives that determine its long-term direction and short-term goals." See Harmon and Jacobs, *The Vital Difference: Unleashing the Powers of Sustained Corporate Success* (New York: AMACOM, 1985), p. 34. In addition, Robert Quinn and M. R. McGrath argue that organizational culture and individual character are "analogues of one another," particularly in the processing and utilization of information. See Quinn and McGrath "The Transformation of Organizational Cultures: A Competing Values Perspective," in P. J. Frost et al., *Organizational Culture* (Newbury Park, CA: Sage Publications, 1985) p. 329. Harmon and Jacobs equate organizational culture with character. Organizational "character is the organizing and executing will of the corporate personality that releases all its energies, mobilizes all its resources, and coordinates all its actions to fulfill its mission and objectives in accordance with its central beliefs and values." See Harmon and Jacobs, p. 91.

3. The same content and process found in the parallel between culture and character is also fundamental to individual character formation. See C. G. Jung, *Psychological Types* (London: Routledge & Kegan Paul, 1923); and *The Integration of Personality* (New York: Farrar and Rinehart, 1939).

4. There is considerable precedent for this conclusion in the history of personality theory and research. In psychoanalysis, Sigmund Freud's *An Outline of Psychoanalysis* (New York: Norton, 1949) and the works of Carl Jung take this perspective. Allport emphasized a person's need to "become" and reach a higher level of wholeness and integrity. See G. W. Allport, *Personality: A Psychological Interpretation* (New York: Holt, 1937) and *Becoming: Basic Considerations for a Psychology of Personality* (New Haven: Yale University Press, 1955). As you have read, Maslow spent many years researching this and related notions. See A. H. Maslow's *Toward a Psychology of Being* (Princeton, NJ: D. Van Nostrand, 1968) and *The Farther Reaches of Human Nature* (New York: Viking, 1971). Others who have taken this perspective include A. Angyal, *Foundations for a Science of Personality* (New York: Commonwealth Fund, 1941); K. Goldstein, *The Organism* (New York:

American Book, 1939); and C. R. Rogers, *Client-centered Therapy* (Boston: Houghton Mifflin, 1951). Roger Harrison offers a definition of organizational effectiveness similar to that presented in this book: "The essence of organizational health is *cultural balance,* a condition in which the tendency toward excess that characterizes any 'pure' organizational culture is mitigated by the opposing tendencies of the other cultures." See R. Harrison, "Harnessing Personal Energy: How Companies Can Inspire Employees," *Organizational Dynamics* (Autumn 1987), p. 13.

5. J. O'Toole, *Vanguard Management: Redesigning the Corporate Future* (Garden City, NY: Doubleday, 1985), p. 136.

6. The most notable example is Tom Peters. Peters and his consultants have much to say that is of considerable value, especially for working managers, but they also imply that the competence culture is the right way for everyone. See T. J. Peters and R. H. Waterman, Jr., *In Search of Excellence: Lessons from America's Best-Run Companies* (New York: Harper & Row, 1982); T. J. Peters and N. Austin, *A Passion for Excellence: The Leadership Difference* (New York: Random House, 1985); and T. J. Peters, *Thriving on Chaos: A Handbook for a Management Revolution* (New York: Alfred A. Knopf, 1987); see also W. G. Ouchi, *Theory Z* (Reading, MA: Addison-Wesley, 1981); and M. R. Weisbord, *Productive Workplaces: Organizing and Managing for Dignity, Meaning, and Community* (San Francisco: Jossey-Bass, 1987).

7. M. S. Peck, *The Road Less Traveled: A New Psychology of Love, Traditional Values and Spiritual Growth* (New York: Simon and Schuster, 1978), p. 51.

8. Quoted in R. Benfari, *Understanding Your Management Style: Beyond the Myers-Briggs Type Indicators* (Lexington, MA: Lexington Books, 1991), p. 131.

9. A. L. Wilkins and N. J. Bristow, "For Successful Organizational Culture, Honor Your Past," *Academy of Management Executive*, 1, no. 3 (1987), pp. 221–28. See also A. L. Wilkins, *Developing Corporate Character: How to Successfully Change An Organization Without Destroying It* (San Francisco: Jossey-Bass, 1989).

10. John P. Kotter and James L. Heskett, *Corporate Culture and Performance* (New York: Free Press, 1992), p. 149.

11. P. F. Drucker, "Don't Change Corporate Culture—Use it!" *Wall Street Journal* (March 28, 1991).

12. F. G. Harmon and G. Jacobs, *The Vital Difference: Unleashing the Powers of Sustained Corporate Success* (New York: AMACOM, 1985), p. 42.

13. D. R. Denison, *Corporate Culture and Organizational Effectiveness.* (New York: John Wiley & Sons, 1990), pp. 189–90.

14. M. Stahlman, "The Failure of IBM: Lessons For The Future." *Upside* (March 1993), pp. 29–30, 34, 37.

15. The mainframe computer industry corresponds to the control culture, the minicomputer industry corresponds to the competence culture, the personal computer industry corresponds to the cultivation culture, and the workstation computer industry corresponds to the collaborative culture.

Corroboration from the Work of Others

A number of experts from various disciplines have identified organizational forms that corroborate the concepts of the four core cultures put forth in this book. A summary of their work follows.

In 1972, Harrison proposed that four "organization ideologies" exist across all organizations. By *ideology* he means "a system of thought that is a central determinant of the character of the organization." His full definition of *organization ideology* comes very close to my definition of organizational culture. Four years later Handy elaborated on Harrison's ideologies and characters, but renamed them "organizational cultures"[1]; in a later publication, Harrison uses the word "culture" as his central organizing concept.

Harrison writes that "these ideologies are seldom found in organizations as pure types, but most organizations tend to center on one or another of them." The four organization ideologies are power orientation, role orientation, task orientation, and person orientation. The power orientation organization attempts "to dominate its environment . . . strive[s] to maintain control over subordinates . . . and always attempts to bargain to [its] own advantage." The role orientation organization emphasizes "hierarchy and status . . . rules and procedures, legitimacy and responsibility . . . and leaves the customer, the public, or the client with little alternate choice in dealing with [it]." The task orientation organization focuses on the "achievement of a superordinate goal as its highest value . . . authority is considered legitimate only if it is based on appropriate knowledge and competence." Finally, the person orientation organization exists "to serve the needs of its members [and] people are generally not expected to do things that are incongruent with their own goals and values."[2] Fifteen years later, Harrison had partially revised his thinking and renamed the task orientation organization the "achievement culture" and the person orientation organization the "support culture." He also pointed to the presence of a "dominant culture [in any one organization] that tends to permeate all of its parts."[3] The connection between the concepts in this book and Harrison's "organization ideologies" is evident. His power orientation and role orientation are closely allied to the control culture; the task/achievement orientation is reminiscent of the competence culture; and the person/support orientation is very similar to the cultivation culture. Harrison does not identify an organization ideology that corresponds to the collaboration culture.

Deal and Kennedy also proposed four "generic cultures" that are formed from the "broader social and business environment in which the company operates." The first culture is the "tough-guy, macho" culture, a "world of individualists who regularly take high risks and get quick feedback on whether their actions were right or wrong." In that culture, internal competition is strong as individuals strive to be stars. Belonging to a team offers them no rewards. This culture has clear parallels with the competence culture. Deal and Kennedy's second culture is the "work hard/play hard" culture in which "fun and action are the rule." This culture is based on the team, not the individual, and it places great value on customers. This culture is reminiscent of the collaboration culture. Their third culture is the "bet-your-company" culture, one of "big-stakes decisions, where years pass before employees know whether decisions have paid off." Rank and respect for authority are important in this culture that parallels the control culture. Finally, Deal and Kennedy describe the "process" culture, which also parallels the control culture, and where there is "little or no feedback" and "employees find it hard to measure what they do." In this hierarchical culture, people focus on the system rather than the individual, and life centers on work patterns and procedures. Deal and Kennedy do not develop a culture that resembles the cultivation culture.[4]

Bennis and Nanus discuss the presence of three "social architectures" that provide meaning and set patterns of behavior for an organization's members. The three social architectures are the formalistic, collegial, and personalistic. The formalistic culture "has a formal structure that emphasizes clear-cut, explicit rules; a formal committee structure; and a clear division of labor into 'finance' and 'operations.' " The collegial culture is highly competitive and concerned with excellence. Yet it encourages participation in decision making to arrive at consensus on all issues. Bennis and Nanus see the root of the collegial culture in the worlds of academia and science. The personalistic culture is flexible. Each individual participates in decision making, with the goal of achieving self-actualization.[5] Again, the parallels with the four core cultures are readily apparent. Bennis and Nanus's formalistic culture is closely allied with the control culture, their personalistic culture corresponds to the cultivation culture, and their collegial culture appears to be a combination of the competence and collaboration cultures.

Hirsh identified four organizational "preferences": (1) the ST organization which emphasizes specific factual details; control and certainty; work roles rather than the worker; and organizational goals and hierarchy; (2) the SF organization which emphasizes specific factual details; the interpersonal features of the work environment; familiarity; and workers first, then work and work roles; (3) the NT organization which emphasizes broad, global concepts; macro issues; efficiency; and an impersonal atmosphere; and the (4) NF organization which emphasizes growth; macro issues, personal and human goals of the organization; and the personal and social needs of people in the organization.[6] Hirsh's organizational preferences closely mirror the four core cultures.

In 1985, O'Toole took a different approach to culture formation. He claims that the thousands of theories of work organization and employee motivation can be placed under one of four broad headings: meritocracy, egalitarianism, behaviorism, and humanism. A meritocracy holds to the belief that individuals have fundamentally different talents and abilities that should be acknowledged. The most productive employees get the richest rewards. According to O'Toole, meritocracy fits with free-market principles operating in the United States. A meritocratic organization coincides closely with the competence culture. Egalitarianism holds that "the common bond of humanity entitles all workers to certain minimal rights." Egalitarian forms of organization are supported by labor unions and socialist economists. Egalitarianism fits closely with the collaboration culture. Behaviorism places order and survival for the group foremost. To ensure this, it closely monitors and regulates employee performance. Here is a tie to the control culture. Finally, humanism holds that "work should be an ennobling experience and that the real purpose of work is to draw out the full potential of all employees."[7] Humanism coincides with the cultivation culture.

The perspectives of Robert Quinn and his colleagues tie directly into the concept of the powerful and pervasive effect of culture in an organization, as presented in this book. They define an organizational culture as a "collective belief system about social arrangements." Culture "involves beliefs about the 'appropriate' nature of transactions." This relates directly to my definition of culture as "how we do things around here in order to succeed." Quinn argues that these beliefs "tend to be deeply embedded values . . . and include [beliefs] about . . . organizational purpose, criteria of performance, the location of authority, legitimate bases of power, decision-making orientations, style of leadership, compliance, evaluation, and motivation."

Quinn and his colleagues recognize four organizational "forms" or cultures: the hierarchical culture, the consensual culture, the rational culture, and the ideological culture. The hierarchical culture, which corresponds to the control culture, has security as its motive. It emphasizes authority, regulations, stability, and control. Its power base is technical knowledge. It makes decisions through factual analysis. Its leaders are conservative and cautious. It evaluates its people by formal criteria. The consensual culture, which corresponds to the collaboration culture, centers on the group and has affiliation as its motive. Its power base is informal status. It makes decisions through group participation. Its leaders are concerned and supportive. And it evaluates its people on the quality of the relationship. The rational culture, which mirrors the competence culture, has achievement as its motive. It emphasizes productivity and efficiency. Its power base is competence. Its decisions occur as pronouncements. Its leaders are directive and goal oriented. It evaluates its people on their productiveness. Finally, the ideological culture, which is similar to the cultivation culture, has growth as its motive. It emphasizes broad purposes, external support, and the acquisition of resources. Its power base is values. It makes decisions by intuition. Its leaders are inventive and willing to take risks. It evaluates its people on the intensity of their effort.[8] While differences occur in a number of subareas, I am in almost total agreement with Quinn and his colleagues on the nature of these four cultures.

Michael Treacy and Fred Wiersema claim that there are three paths to market leadership: operational excellence, customer intimacy, and product leadership. Operational excellence means "providing customers with reliable products or services at competitive prices and delivered with minimal difficulty or inconvenience." Companies that focus on operational excellence strive to cut costs in a number of ways—on overhead, the elimination of intermediate production steps, and on transactions and other "friction" costs. At the same time they work to create the best business processes at all functions and across the organization. This type of market leadership corresponds closely to that of the control culture. Customer intimacy means "segmenting and targeting markets precisely and then tailoring offerings to match exactly the demands of those niches. Companies that excel in customer intimacy combine detailed customer knowledge with operational flexibility so they can respond quickly to almost any need . . . As a consequence, these companies engender tremendous customer loyalty." This path is closely reminiscent of the collaboration culture. Product leadership means "offering customers leading-edge products and services that consistently embrace the customer's use or application of the product, thereby making rivals' goods obsolete." By producing state-of-the-art products and services, this path mirrors the competence culture. Treacy and Wiersema do not discuss a fourth path that might relate to the cultivation culture. However, they do place great emphasis on the importance of establishing a primary focus (e.g., market leadership) and staying with it. "The greater challenge is to sustain that focus, to drive that strategy relentlessly through the organization, to develop the internal consistency, and to confront radical change."[9]

NOTES

1. C. B. Handy, *Understanding Organizations* (New York: Penguin Books, 1976).

2. R. Harrison, "Understanding Your Organization's Character," *Harvard Business Review*, 3 (1972), pp. 119–28.

3. ———. "Harnessing Personal Energy: How Companies Can Inspire Employees," *Organizational Dynamics* (Autumn 1987), pp. 4–20.

4. T. E. Deal and A. A. Kennedy, *Corporate Cultures: The Rites and Rituals of Corporate Life* (Reading, MA: Addison-Wesley, 1982).

5. W. Bennis and B. Nanus, *Leaders: The Strategies For Taking Charge* (New York: Harper & Row, 1985).

6. S. Hirsh, *Using the Myers-Briggs Type Indicator in Organizations: A Resource Book* (Palo Alto, CA: Consulting Psychologists Press, 1985).

7. J. O'Toole, *Vanguard Management: Redesigning the Corporate Future* (Garden City, NY: Doubleday, 1985).

8. J. F. Kimberly and R. E. Quinn, *Managing Organizational Transitions* (Homewood, IL: Irwin, 1984); R. E. Quinn and M. R. McGrath, (1985) "The Transformation of Organizational Cultures: A Competing Values Perspective," in *Organizational Culture*, ed. P. J. Frost et al. (Beverly Hills, CA: Sage Publications, 1985); R. E. Quinn, *Beyond Rational Management: Mastering the Paradoxes and Competing Demands of High Performance* (San Francisco: Jossey-Bass, 1988).

9. M. Treacy and F. Wiersema, "Customer Intimacy and Other Value Disciplines," *Harvard Business Review* (January–February 1993), pp. 84–93.

References

Adams, G. B., and Ingersoll, V. H. "Culture, Technical Rationality, and Organizational Culture." *American Review of Public Administration,* 20, no. 4 (December 1990): pp. 285–302.

Allaire, Y., and Firsirotu, M. E. "Theories of Organizational Culture." *Organizational Studies,* 5, no. 3 (1984): pp. 193–226.

Allport, G. W. *Becoming: Basic Considerations for a Psychology of Personality.* New Haven: Yale University Press, 1955.

──────. *Personality: A Psychological Interpretation.* New York: Holt, 1937.

American Bankers Association and Ernst & Whinney. *Implementing Mergers and Acquisitions in the Financial Services Industry: From Handshake to Hands On.* Washington, DC: American Bankers Association, 1985.

Angyal, A. *Foundations for a Science of Personality.* New York: Commonwealth Fund, 1941.

Anthony, P. D. "The Paradox of the Management of Culture or 'He Who Leads Is Lost.' " *Personnel Review,* 19, no. 4 (1990): pp. 3–8.

Argyris, C. "Teaching Smart People How to Learn." *Harvard Business Review* (May–June 1991): pp. 99–109.

Baker, E. L. "Managing Organizational Culture." *Management Review* (July 1980): pp. 8–13.

Barley, S.; Meyer, G. W.; & Gash, D. "Cultures of Culture: Academics, Practitioners, and the Pragmatics of Normative Control." *Administrative Science Quarterly,* 33 (1988): pp. 24–60.

Barnouw, V. *Culture and Personality.* Homewood, IL: Dorsey Press, 1979.

Baum, H. S. "Organizational Politics Against Organizational Culture: A Psychoanalytic Perspective." *Human Resource Management,* 28, no. 2 (Summer 1989): pp. 191–206.

Benfari, R. *Understanding Your Management Style: Beyond the Myers-Briggs Type Indicator.* Lexington, MA: Lexington Books, 1991.

Bennigson, L. A. "Managing Corporate Cultures." *Management Review* (February 1985): pp. 31–32.

Bennis, W. *Why Leaders Can't Lead: The Unconscious Conspiracy Continues.* San Francisco: Jossey-Bass, 1990.

Bennis, W., and Nanus, B. *Leaders: The Strategies For Taking Charge.* New York: Harper & Row, 1985.

Berger, P., and Luckmann, T. *The Social Construction of Reality: A Treatise in the Sociology of Knowledge.* Garden City, NY: Doubleday, 1966.

Botterill, M. "Changing Corporate Culture." *Management Services*, 34, no. 6 (June 1990): pp. 14–18.

Bower, M. *The Will To Manage.* New York: McGraw-Hill, 1966.

Bowers, B. "Follow the Leader: Recruiting Employees Who Share the Corporate Vision Requires a Delicate Balancing Act." *Wall Street Journal Reports* (November 22, 1991): p. R16.

Bridges, W. *The Character of Organizations: Using Jungian Types in Organizational Development.* Palo Alto, CA: Consulting Psychologists Press, 1992.

Buday, R. S. "Forging A New Culture At Capital Holding's Direct Response Group." *Insights Quarterly*, 4, no. 2 (Fall 1992): pp. 38–49.

Buono, A. F., and Bowditch, J. L. *The Human Side of Mergers and Acquisitions: Managing Collisions Between People, Cultures, and Organizations.* San Francisco: Jossey-Bass, 1989.

Burke, W. "Organizational Culture." *Organizational Dynamics*, 12, no. 2 (1983): pp. 4–80.

Case, J. "Collective Effort." *Inc.* (January 1992): pp. 32–43.

Clark, R. *The Distinctive College: Antioch, Reed, and Swarthmore.* New York: Aldine, 1970.

Clifford, D. K., and Cavanagh, R. E. *The Winning Performance: How America's High-Growth Midsize Companies Succeed.* New York: Bantam Books, 1985.

Cole, M., and Schribner, S. *Culture and Thought.* New York: John Wiley & Sons, 1974.

Covey, S. R. *Principle-Centered Leadership.* New York: Summit Books, 1991.

————. *The Seven Habits of Highly Effective People: Restoring the Character Ethic.* New York: Simon & Schuster, 1989.

Davis, S. M. *Future Perfect.* Reading, MA: Addison-Wesley, 1987.

————. *Managing Corporate Culture.* Cambridge, MA: Ballinger, 1984.

Deal, T. E., and Kennedy, A. A. *Corporate Cultures: The Rites and Rituals of Corporate Life.* Reading, MA: Addison-Wesley, 1982.

DeLisi, P. S. "Lessons from the Steel Axe: Culture, Technology, and Organizational Change." *Sloan Management Review*, 32, no. 1 (Fall 1990): pp. 83–93.

Denison, D. R. *Corporate Culture and Organizational Effectiveness*. New York: John Wiley & Sons, 1990.

DePree, M. *Leadership Is An Art*. New York: Bantam Doubleday Dell, 1989.

Driver, M. J., and Rowe, A. J. "Decision-making Styles: A New Approach to Management Decision Making." In *Behavioral Problems in Organizations*. Edited by C. Cooper. Englewood Cliffs, NJ: Prentice-Hall, 1979.

Drucker, P. F. *The Changing World of the Executive*. New York: Times Books, 1982.

————. "Don't Change Corporate Culture—Use It!" *The Wall Street Journal* (March 28, 1991).

————. *Innovation and Entrepreneurship: Practice and Principles*. New York: Harper & Row, 1985.

————. *Management: Tasks, Responsibilities, Practices*. New York: Harper & Row, 1973.

————. *Managing for Results*. New York: Harper & Row, 1964.

————. *Managing In Turbulent Times*. New York: Harper & Row, 1980.

————. *The Practice of Management*. New York: Harper & Row, 1954.

Duncan, W. J. "Organizational Culture: 'Getting a Fix' on an Elusive Concept." *The Academy of Management Executive*, 3, no. 3 (August 1989): pp. 229–236.

Fayol, H. *General and Industrial Administration*. New York: Pitman, 1949.

Feldman, S. P. "How Organizational Culture Can Affect Innovation." *Organizational Dynamics* (Summer 1988): pp. 57–68.

Fitzgerald, T. H. "Can Change in Organizational Culture Really Be Managed?" *Organizational Dynamics* (Autumn 1988): pp. 4–15.

Forgus, R., and Shulman, B. H. *Personality: A Cognitive View*. Englewood Cliffs, NJ: Prentice-Hall, 1979.

Freud, S. *An Outline of Psychoanalysis*. New York: Norton, 1949.

Frost, P. J.; Moore, L. F.; Louis, M. L.; Lundberg, C. C.; and Martin, J., eds. *Organizational Culture*. Newbury Park, CA: Sage, 1985.

————. *Reframing Organizational Culture*. Newbury Park, CA: Sage, 1991.

Fuchsberg, G. "Quality Programs Show Shoddy Results." *The Wall Street Journal* (May 14, 1992).

Garsombke, D. J. "Organizational Culture Dons the Mantle of Militarism." *Organizational Dynamics*, 17, no. 1 (Summer 1988): pp. 46–56.

Geertz, C. *The Interpretation of Cultures.* New York: Basic Books, 1973.

Goldstein, K. *The Organism.* New York: American Book, 1939.

Goodall, H. L. *Casing a Promised Land: The Autobiography of an Organizational Detective as Cultural Ethnographer.* Carbondale, IL: Southern Illinois University Press, 1989.

Gordon, G. G. "Industry Determinants of Organizational Culture." *Academy of Management Review,* 16, no. 2 (April 1991): pp. 396–415.

Gregory, K. L. "Native-view Paradigms: Multiple Cultures and Culture Conflicts in Organizations." *Administrative Science Quarterly,* 28, no. 3 (September 1983): pp. 359–76.

Gross, J. L., and Rayner, S. *Measuring Culture: A Paradigm for the Analysis of Social Organization.* New York: Columbia University Press, 1985.

Gummer, B. "Managing Organizational Cultures: Management Science or Management Ideology?" *Administration in Social Work,* 14, no. 1 (1990): pp. 135–53.

Gupta, U. "Keeping the Faith: First, There's the Entrepreneur's Vision. Then Success. But Can the Vision Survive?" *The Wall Street Journal Reports* (November 22, 1991): p. R16.

Hall, C. S., and Lindzey, G. *Theories of Personality.* New York: John Wiley & Sons, 1957.

Hall, E. T. *Beyond Culture.* New York: Anchor, 1976.

Hampden-Turner, C. *Creating Corporate Culture: From Discord to Harmony.* Reading, MA.: Addison-Wesley, 1990.

——— . *Maps of the Mind.* New York: Macmillan, 1981.

Handy, C. B. *Understanding Organizations.* New York: Penguin Books, 1976.

Hannabuss, S. "The Management of Meaning." *Management Decision,* 27, no. 6 (1989): pp. 16–21.

Hannon, M. T., and Freeman, J. *Organizational Ecology.* Cambridge: Harvard University Press, 1989.

Harmon, F. G., and Jacobs, G. *The Vital Difference: Unleashing the Powers of Sustained Corporate Success.* New York: AMACOM, 1985.

Harrigan, K. R. *Strategies for Joint Ventures.* Lexington, MA: Lexington Books, 1985.

Harris, P. R., and Moran, R. T. *Managing Cultural Differences.* Houston: Gulf, 1979.

Harrison, R. "Diagnosing Organization Ideology." In *Annual Handbook for Group Facilitators.* Edited by J. E. Jones and J. W. Pfeiffer. La Jolla, CA: University Associate Publishers, 1975.

————. "Harnessing Personal Energy: How Companies Can Inspire Employees." *Organizational Dynamics* (Autumn 1987): pp. 4–20.

————. "Understanding Your Organization's Character." *Harvard Business Review*, 3 (1972): pp. 119–28.

Hassard, J., and Sharifi, S. "Corporate Culture and Strategic Change." *Journal of General Management*, 15, no. 2 (Winter 1989): pp. 4–19.

Hebden, J. E. "Adopting an Organization's Culture: The Socialization of Graduate Trainees." *Organizational Dynamics* (Summer 1986): pp. 54–72.

Hennestad, B. W. "The Symbolic Impact of Double Bind Leadership: Double Bind and the Dynamics of Organizational Culture." *Journal of Management Studies*, 27, no. 3 (May 1990): pp. 265–80.

Hickman, C. R., and Silva, M. A. *Creating Excellence: Managing Corporate Culture, Strategy, and Change in the New Age.* New York: New American Library, 1984.

Hirsh, S. *Using the Myers-Briggs Type Indicator in Organizations.* Palo Alto, CA: Consulting Psychologists Press, 1985.

Hirsh, S., and Kummerow, J. *Introduction to Type in Organizations: Individual Interpretive Guide.* Palo Alto, CA: Consulting Psychologists Press, 1990.

————. *Lifetypes.* New York: Warner Books, 1989.

Hofstede, G. *Culture's Consequences.* Newbury Park, CA: Sage, 1980.

Hofstede, G.; Neuijen, B.; Ohayv, D. D.; and Sanders, G. "Measuring Organizational Cultures: A Qualitative and Quantitative Study Across Twenty Cases." *Administrative Science Quarterly*, 35, no. 2 (June 1990): pp. 286–316.

Horton, T. R. *What Works For Me: 16 CEOs Talk About Their Careers and Commitments.* New York: Random House, 1986.

Hrebiniak, L. G. "The Organizational Environmental Research Program: An Overview and Critique." In *Perspectives on Organization Design and Behavior.* Edited by A. H. Van de Ven & W. F. Joyce. New York: John Wiley & Sons, 1981.

Hsu, F. L. K. *Psychological Anthropology.* Cambridge, MA: Schenkman, 1972.

Hymowitz, C. "Which Corporate Culture Fits You?" *The Wall Street Journal* (July 17, 1989), p. B1.

Jaccaci, A. T. "The Social Architecture of a Learning Culture." *Training & Development Journal*, 43, no. 11 (November 1989): pp. 49–51.

James, G. "State-of-the-Art Technology and Organisational Culture." *Management Decision*, 29, no. 2 (1991): pp. 18–31.

Jaques, E. *The Changing Culture of a Factory*. New York: Dryden Press, 1951.

————. *Requisite Organization: The CEO's Guide to Creative Structure and Leadership*. Arlington, VA: Cason Hall, 1989.

Joiner, C. W., Jr. *Leadership For Change*. Cambridge, MA: Ballinger, 1987.

Jones, W. T. *The Romantic Syndrome: Towards a New Method in Cultural Anthropology and the History of Ideas*. The Hague: Martinus Nijhoff, 1961.

Jung, C. G. *Analytical Psychology*. New York: Moffat, Yard, 1916.

————. *Collected Works: The Development of Personality*. Vol. 17. New York: Pantheon Press, 1954.

————. *The Integration of Personality*. New York: Farrar and Rinehart, 1939.

————. *Modern Man In Search of a Soul*. New York: Harcourt, 1933.

————. *Psychological Types*. London: Routledge & Kegan Paul, 1923.

————. *Psychology of the Unconscious*. New York: Dodd, 1925.

————. *The Theory of Psychoanalysis*. New York: Nervous and Mental Disease Publishing, 1915.

Keen, P. G. W. "Cognitive Style and Career Specialization." In *Organizational Careers: Some New Perspectives*. Edited by J. Van Maanen. New York: John Wiley & Sons, 1977.

Kets de Vries, M. F., and Miller, D. "Personality, Culture, and Organization." *Academy of Management Review*, 11, no. 2 (April 1986): pp. 266–79.

Kilmann, R. H. "Corporate Culture: Managing the Intangible Style of Corporate Life May Be the Key to Avoiding Stagnation." *Psychology Today* (April 1985): pp. 62–68.

————. "A Typology of Organization Typologies: Toward Parsimony and Integration." *Human Relations*, 36 (1983): pp. 523–48.

Kilmann, R. H.; Saxton, M. J.; Serpa, R. & Associates. *Gaining Control of the Corporate Culture*. San Francisco: Jossey-Bass, 1985.

Kimberly, J. F., and Quinn, R. E. *Managing Organizational Transitions*. Homewood, IL: Richard D. Irwin, 1984.

Kindel, S. "The Bundle Book. At Reuben Mark's Colgate, Attention to Small Details Creates Large Profits." *Financial World* (January 5, 1993): pp. 34–35.

Kirp, D. L., and Rice, D. S. "Fast Forward—Styles of California Management." *Harvard Business Review*, 1 (January–February 1988): pp. 74–83.

Kluckhohn, F. R., and Strodtbeck, F. L. *Variations in Value Orientations.* New York: Harper & Row, 1961.

Knight, C. F. "Emerson Electric: Consistent Profits, Consistently." *Harvard Business Review,* (January–February 1992): pp. 57–70.

Kotter, J. P., and Heskett, J. L. *Corporate Culture and Performance.* New York: Free Press, 1992.

Kouzes, J. M., and Posner, B. Z. *The Leadership Challenge: How to Get Extraordinary Things Done in Organizations.* San Francisco: Jossey-Bass, 1987.

Kuhn, T. S. *The Structure of Scientific Revolutions.* 2d ed. Chicago: University of Chicago Press, 1970.

Lawrence, P. "The Harvard Organization and Environment Research Program." In *Perspectives on Organization Design and Behavior.* Edited by A. H. Van de Ven and W. F. Joyce. New York: John Wiley & Sons, 1981.

Lawrence, P., and Lorsch, J. *Organizations and Environment.* Cambridge: Harvard University Press, 1967.

Leinberger, P., and Tucker, B. *The New Individualists: The Generation After The Organization Man.* New York: HarperCollins, 1991.

Levering, R. *A Great Place To Work: What Makes Some Employers So Good (and Most So Bad).* New York: Random House, 1988.

Levering, R., and Moskowitz, M. *The 100 Best Companies To Work For In America.* New York: Currency/Doubleday, 1993.

Levering, R.; Moskowitz, M.; and Katz, M. *The 100 Best Companies To Work For In America.* Reading, MA: Addison-Wesley, 1984.

Levinson, H., and Rosenthal, S. *CEO: Corporate Leadership in Action.* New York: Basic Books, 1984.

Lincoln, J. R., and Kalleberg, A. L. *Culture, Control, and Commitment: A Study of Work Organization and Work Attitudes in the United States and Japan.* Cambridge: Cambridge University Press, 1990.

Louis, M. R. "A Cultural Perspective on Organizations." *Human Systems Management,* 2 (1981): pp. 246–58.

―――― . "Organizations as Culture Bearing Milieux." In *Organizational Symbolism.* Edited by L. R. Pondy et al. Greenwich, CT: JAI Press, 1983.

Lundberg, C. C. "Innovative Organisation Development Practices: Part II—Surfacing Organisational Culture." *Journal of Managerial Psychology,* 5, no. 4 (1990): pp. 19–26.

Maddi, S. R. *Personality Theories: A Comparative Analysis.* Homewood, IL: Dorsey Press, 1968.

Margulies, W. "Make the Most of Your Corporate Identity." *Harvard Business Review*, 55 (July–August 1977): p. 184.

Markley, M. "Need for Control Undermined Raymond." *Houston Chronicle* (June 9, 1991).

Marks, M., and Mirvis, P. H. "The Merger Syndrome: When Corporate Cultures Clash." *Psychology Today* (October 1986): pp. 36–42.

Maslow, A. H. *The Farther Reaches of Human Nature*. New York: Viking Press, 1971.

———. *Motivation and Personality*. New York: Harper, 1954.

———. *Toward a Psychology of Being*. Princeton, NJ: D. Van Nostrand, 1968.

McClelland, D. C. *The Achieving Society*. Princeton, NJ: D. Van Nostrand, 1961.

———. *Power: The Inner Experience*. New York: Irvington, 1975.

McClelland, D. C., and Burnham, D. H. "Power Is the Great Motivator." *Harvard Business Review* (March–April, 1976): p. 100.

McClelland, D. C., and Winters, D. G. *Motivating Economic Achievement*. New York: Free Press, 1969.

Meyer, G. "Review: Assessing Diversity and Culture on the PC." *HRMagazine*, 36, no. 4 (April 1991): pp. 25–28.

Mills, D. Q. *Rebirth of the Corporation*. New York: John Wiley & Sons, 1991.

Mitroff, I. I. "Archetypal Social Systems Analysis: On the Deeper Structure of Human Systems." *Academy of Management Review*, 8 (1983): pp. 387–97.

———. *Stakeholders of the Organizational Mind*. San Francisco: Jossey-Bass, 1983.

Mitroff, I. I., and Mason, R. O. "Business Policy and Metaphysics: Some Philosophical Considerations." *Academy of Management Review*, 7, (1982): pp. 361–70.

Moeller, P. "What Counts in Corporate Life." *Houston Chronicle* (September 29, 1983).

Morgan, G. *Images of Organizations*. Newbury Park, CA: Sage, 1986.

Myers, I. B. *Introduction To Type*. Rev. ed. Palo Alto, CA: Consulting Psychologists Press, 1987.

Myers, I. B., and Myers, P. B. *Gifts Differing*. Palo Alto, CA: Consulting Psychologists Press, 1980.

Myers, L. "Borland Finding Success High Above Silicon Valley." *Houston Chronicle* (March 8, 1992).

Nadler, D. A.; Gerstein, M. S.; Shaw, R. B. & Associates. *Organizational Architecture: Designs for Changing Organizations*. San Francisco, CA: Jossey-Bass, 1992.

Nichols, N. A. "Profits With a Purpose: An Interview With Tom Chapman." *Harvard Business Review* (November–December 1992): pp. 86–95.

O'Boyle, T. F., and Russel, M. "Troubled Marriage: 'Steel Giants' Merger Brings Headaches, J&L and Republic Find." *Wall Street Journal* (November 30, 1984).

Odom, R. Y.; Boxx, W. R., and Dunn, M. G. "Organizational Cultures, Commitment, Satisfaction, and Cohesion." *Public Productivity & Management Review*, 14, no. 2 (Winter 1990): pp. 157–69.

O'Toole, J. J. "Corporate and Managerial Cultures." In *Behavioral Problems in Organizations*. Edited by C. L. Cooper. Englewood Cliffs, NJ: Prentice-Hall, 1979.

O'Toole, J. *Vanguard Management: Redesigning the Corporate Future*. Garden City, NY: Doubleday, 1985.

Ouchi, W. G. *Theory Z*. Reading, MA: Addison-Wesley, 1981.

Ouchi, W. G., and Wilkins, A. L. "Organizational Culture." *Annual Review of Sociology*, 11 (1985): pp. 457–83.

Pascale, R. T. "Fitting New Employees into the Company Culture." *Fortune* (May 28, 1984): pp. 28–40.

——— . *Managing on the Edge: How the Smartest Companies Use Conflict to Stay Ahead*. New York: Simon & Schuster, 1990.

——— . "The Paradox of 'Corporate Culture': Reconciling Ourselves to Socialization." *California Management Review*, 13, no. 4 (1985): pp. 546–58.

Payne, R. "Taking Stock of Corporate Culture." *Personnel Management*, 23, no. 7 (July 1991): pp. 26–29.

Peck, M. S. *The Road Less Traveled: A New Psychology of Love, Traditional Values and Spiritual Growth*. New York: Simon & Schuster, 1978.

Peters, T. J. *Liberation Management: Necessary Disorganization for the Nanosecond Nineties*. New York: Alfred A. Knopf, 1992.

——— . *Thriving On Chaos: A Handbook for a Management Revolution*. Alfred A. Knopf, 1987.

Peters, T. J., and Austin, N. *A Passion for Excellence: The Leadership Difference*. New York: Random House, 1985.

Peters, T. J., and Waterman, R. H., Jr. *In Search of Excellence: Lessons from America's Best-Run Companies*. New York: Harper & Row, 1982.

Pettigrew, A. M. "On Studying Organizational Cultures." *Administrative Science Quarterly*, 24 (1979): pp. 570–81.

Poupart, R., and Hobbs, B. "Changing the Corporate Culture to Ensure Success: A Practical Guide." *National Productivity Review*, 8, no. 3 (Summer 1989): pp. 223–38.

Pritchett, P. *After the Merger: Managing the Shockwaves*. Homewood, IL: Dow Jones-Irwin, 1985.

Pulley, B. "Impulse Item: Wrigley Is Thriving, Despite the Recession, in a Resilient Business." *Wall Street Journal* (May 29, 1991).

Quinn, R. E. *Beyond Rational Management: Mastering the Paradoxes and Competing Demands of High Performance*. San Francisco: Jossey-Bass, 1988.

Quinn, R. E., and Cameron, K. S. *Paradox and Transformation: Toward A Theory of Change in Organization and Management*. Cambridge, MA: Ballinger, 1988.

Quinn, R. E.; Faerman, S. R.; Thompson, M. P.; and McGrath, M. R. *Becoming A Master Manager: A Competency Framework*. New York: John Wiley & Sons, 1990.

Quinn, R. E., and Hall, R. H. "Environments, Organizations, and Policy Makers: Towards an Integrative Framework." In *Organization Theory and Public Policy*. Edited by R. H. Hall and R. E. Quinn. Newbury Park, CA: Sage, 1983.

Quinn, R. E., and McGrath, M. R. "The Transformation of Organizational Cultures: A Competing Values Perspective." In *Organizational Culture*. Edited by P. J. Frost et al. Newbury Park, CA: Sage, 1985.

Ray, M., and Rinzler, A., eds. *The New Paradigm in Business: Emerging Strategies for Leadership and Organizational Change*. New York: Jeremy P. Tarcher/Perigee Books, 1993.

Reynolds, P. C. "Imposing a Corporate Culture." *Psychology Today* (March 1987): pp. 33–38.

Rifkin, G. "The Hand That Shakes Digital Up." *New York Times* (May 15, 1992): pp. C1, C5.

Ritti, R. R., and Funkhouser, G. R. *The Ropes To Skip & The Ropes To Know: The Inner Life of an Organization*. New York: John Wiley & Sons, 1987.

Rogers, C. R. *Client-centered Therapy: Its Current Practice, Implications, and Theory*. Boston: Houghton, 1951.

Rosen, M. "Coming to Terms with the Field: Understanding and Doing Organizational Ethnography." *Journal of Management Studies*, 28, no. 1 (January 1991): pp. 1–24.

Rothenberg, A. *The Emerging Goddess: The Creative Process in Art, Science and other Fields*. Chicago: University of Chicago Press, 1979.

Rousseau, D. M. "Normative Beliefs in Fund-Raising Organizations: Linking Culture to Organizational Performance and Individual Responses." *Group & Organization Studies*, 15, no. 4 (December 1990): pp. 448–60.

Roy, R. *The Cultures of Management*. Baltimore: Johns Hopkins University Press, 1970.

Sackmann, S. A. *Cultural Knowledge in Organizations: Exploring the Collective Mind*. Newbury Park, CA: Sage, 1991.

Sathe, V. *Culture and Related Corporate Realities*. Homewood, Ill.: Richard D. Irwin, 1985.

———. "Implications of Corporate Culture: A Manager's Guide to Action." *Organizational Dynamics* (Autumn 1983): pp. 5–23.

———. *Managerial Action and Corporate Culture*. Homewood, Ill.: Richard D. Irwin, 1985.

Schein, L. *A Manager's Guide to Corporate Culture*. New York: The Conference Board, 1989.

———. "Organizational Culture." *American Psychologist*, 45, no. 2 (February 1990): pp. 109–19.

———. *Organizational Culture and Leadership*. San Francisco: Jossey-Bass, 1985.

Schneider, B., ed. *Organizational Climate and Culture*. San Francisco: Jossey-Bass, 1990.

Schwartz, H., and Davis, S. M. "Matching Corporate Culture and Business Strategy." *Organizational Dynamics* (Summer 1981): pp. 30–48.

Scott, M., and Rothman, H. *Companies With a Conscience: Intimate Portraits of Twelve Firms That Make a Difference*. New York: Carol Publishing Group, 1992.

Scott, W. Richard. "Theoretical Perspectives." In *Environments and Organizations*. Edited by M. W. Meyer & Associates. San Francisco: Jossey-Bass, 1978.

Shils, E. *Center and Periphery: Essays in Macrosociology*. Chicago: University of Chicago Press, 1975.

Shils, E. "Center and Periphery: An Idea and its Career, 1935–1987." In *Center and Periphery: An Idea and its Career*. Edited by L. Greenfield & M. Martin. Chicago: University of Chicago Press, 1988.

Shockley-Zalabak, P., and Morley, D. D. "Adhering to Organizational Culture: What Does It Mean: Why Does It Matter?" *Group & Organization Studies*, 14, no. 4 (December 1989): pp. 483–500.

Shull, F. A.; Delbecq, A. L.; and Cummings, L. L. *Organizational Decision Making*. New York: McGraw-Hill, 1970.

Silverzweig, S., and Allen, R. F. "Changing the Corporate Culture." *Sloan Management Review,* 17 (1976): pp. 33–49.

Smircich, L. "Concepts of Culture and Organizational Analysis." *Administrative Science Quarterly,* 28, no. 3 (1983): pp. 339–58.

Sperry, L.; Mickelson, D. J.; and Hunsaker, P. L. *You Can Make It Happen: A Guide to Self-Actualization and Organizational Change.* Reading, Mass.: Addison-Wesley, 1977.

Srivastva, S. & Associates. *The Executive Mind.* San Francisco: Jossey-Bass, 1983.

Stahlman, M. "The Failure of IBM: Lessons For the Future." *Upside* (March 1993): pp. 28–50.

Sutton, C. D., and Nelson, D. L. "Elements of the Cultural Network: The Communicators of Corporate Values." *Leadership & Organization Development Journal,* 11, no. 5 (1990): pp. 3–10.

Swierczek, F. W. "Images of Organisation: Culture and the Management of Technology." *Journal of Managerial Psychology,* 4, no. 3 (1989): pp. 3–10.

Taylor, F. W. *Principles of Scientific Management.* New York: Harper, 1911.

Taylor, W. "The Business of Innovation: An Interview with Paul Cook." *Harvard Business Review,* 2 (March–April 1990): pp. 96–106.

Teal, T. "Service Comes First: An Interview with USAA's Robert F. McDermott." *Harvard Business Review* (September–October 1991): pp. 116–27.

Tichy, N. *The Transformational Leader.* New York: John Wiley & Sons, 1987.

Treacy, M. and Wiersema, F. "Customer Intimacy and Other Value Disciplines." *Harvard Business Review* (January–February 1993): pp. 84–93.

Tregoe, B. B., and Zimmerman, J. W. *Top Management Strategy: What It Is and How to Make It Work.* New York: Simon & Schuster, 1980.

Tucker, R. W.; McCoy, W. J.; and Evans, L. C. (1990). "Can Questionnaires Objectively Assess Organisational Culture?" *Journal of Managerial Psychology,* 5, no. 4 (1990): pp. 4–11.

Ulrich, D., and Lake, D. *Organizational Capability: Competing from the Inside Out.* New York: John Wiley & Sons, 1990.

Van Maanen, J., and Barley, S. R. "Occupational Communities: Culture and Control in Organizations." In *Research in Organizational Behavior.* Edited by B. M. Staw and L. L. Cummings. Vol. 6, Greenwich, CT: JAI Press, 1984.

Van Maanen, J., and Schein, E. H. "Toward a Theory of Organizational Socialization." In *Research in Organizational Behavior.* Edited by B. M. Staw and L. L. Cummings. Vol. 1. Greenwich, CT: JAI Press, 1979.

Walton, M. *The Deming Management Method.* New York: Dodd, Mead, 1986.

Webber, A. M. "Consensus, Continuity, and Common Sense: An Interview with Compaq's Rod Canion." *Harvard Business Review,* 4, (July–August 1990): pp. 114–23.

————. "Crime and Management: An Interview with NYC Police Commissioner Lee P. Brown." *Harvard Business Review* (May–June 1991): pp. 110–26.

Weber, M. *The Theory of Social and Economic Organization.* New York: Oxford University Press, 1947.

Weihrich, H. "How to Achieve Excellence by Managing the Culture in Your Company." *Industrial Management,* 31, no. 5 (September–October 1989): pp. 28–32.

Weinshall, T. D. *Culture and Management.* London: Penguin, 1977.

Weisbord, M. R. *Productive Workplaces: Organizing and Managing for Dignity, Meaning, and Community.* San Francisco: Jossey-Bass, 1987.

Welles, E. O. "Captain Marvel." *Inc.* (January 1992): pp. 44–47.

Wells, K., and Hymowitz, C. "Takeover Trauma: Gulf's Managers Find Merger into Chevron Forces Many Changes." *Wall Street Journal* (December 5, 1984).

White, L. *The Science of Culture.* New York: Grove Press, 1949.

Wick, C. W., and Leon, L. S. *The Learning Edge: How Smart Managers and Smart Companies Stay Ahead.* New York: McGraw Hill, 1993.

Wilkins, A. L. "The Creation of Company Cultures: The Role of Stories and Human Resource Systems." *Human Resources Management,* 23, no. 1 (1984): pp. 41–60.

————. "The Culture Audit: A Tool for Understanding Organizations." *Organizational Dynamics* (Autumn 1983): pp. 24–38.

————. *Developing Corporate Character: How To Successfully Change An Organization Without Destroying It.* San Francisco: Jossey-Bass, 1989.

Wilkins, A. L., and Bristow, N. J. "For Successful Organizational Culture, Honor Your Past." *Academy of Management Executive,* 1, no. 3 (1987): pp. 221–28.

Wilkins, A. L., and Ouchi, W. G. "Efficient Cultures: Exploring the Relationship Between Culture and Organizational Performance." *Administrative Science Quarterly,* 28, no. 3 (1983): pp. 468–81.

Woodman, R. W. "Organizational Change and Development: New Arenas for Inquiry and Action." *Journal of Management*, 15, no. 2 (June 1989): pp. 205–28.

Young, E. "On the Naming of the Rose: Interests and Multiple Meanings as Elements of Organizational Culture." *Organization Studies*, 10, no. 2 (1989): pp. 187–206.

Zachary, G. P. " 'Theocracy of Hackers' Rules Autodesk Inc., A Strangely Run Firm." *Wall Street Journal* (May 28, 1992).

Index